110 Thyristor Projects using S.C.R.s and Triacs

Other Books by R.M. Marston

20 Solid State Projects for the Car and Garage
110 Semiconductor Projects for the Home Constructor
110 Integrated Circuit Projects for the Home Constructor
110 Operational Amplifier Projects for the Home Constructor
110 CMOS Digital IC Projects for the Home Constructor
110 Electronic Alarm Projects for the Home Constructor
110 Waveform Generator Projects for the Home Constructor

110 Thyristor Projects using S.C.R.s and Triacs

R.M. MARSTON

Heinemann : London

William Heinemann Ltd.,
10 Upper Grosvenor Street, London W1X 9PA

LONDON   MELBOURNE   JOHANNESBURG   AUCKLAND

First published by Newnes Technical Books 1972
Reprinted 1975, 1977, 1978, 1981
First published by William Heinemann Ltd., 1986
© R. M. Marston 1972

ISBN 0 434 91215 8

Printed by Billing & Sons Ltd., Worcester.

PREFACE

S.C.R.s and triacs are high-speed solid-state power switches. They can operate at potentials up to hundreds of volts, and can handle currents up to tens or hundreds of amps. They have a multitude of applications in the home and in industry, and can readily be used to control electric lamps, motors, heaters and alarms.

This book is intended to be of equal interest to the electronics amateur, student and engineer. With this aim in mind, the volume starts off by outlining the essential operating characteristics of the s.c.r. and the triac, and then goes on to show 110 useful projects in which these devices can be used. All of these projects have been designed, built and fully evaluated by the author, and range from simple electronic alarms to highly sophisticated self-regulating electric-heater power controllers. Many of the projects use advanced design concepts, and are of outstanding technical interest. The operating principle of each project is explained in concise but comprehensive terms, and brief constructional notes are given where necessary.

All of the projects are designed around internationally available components. The semiconductors used are of American manufacture, but are readily available in all parts of the western world. The volume is specifically designed to be of equal interest to both American and English readers. Where applicable, alternative component values are given for use on both 120 V and 240 V power lines, the 240 V values being noted in parentheses in the appropriate circuit diagrams. The outlines of all semiconductors used in the projects are given in the appendix, as an aid to construction. Unless otherwise stated, all resistors used in the projects are standard half-watt types.

R. M. Marston

CONTENTS

1	Basic Principles and Projects	1
2	15 A.C. Power-Switching Projects	23
3	20 Electronic Alarm Projects	38
4	15 Time-Delay Projects	54
5	25 Lamp-Control Projects	68
6	15 Heater-Control Projects	94
7	15 Universal-Motor Control Projects	110
8	5 Miscellaneous Projects	123
	Appendix	132
	Index	135

CHAPTER 1

BASIC PRINCIPLES AND PROJECTS

S.C.R.s and triacs are members of the thyristor family. They act as high-speed power switches. They are solid-state devices, and can operate at potentials up to several hundred volts, and can handle currents up to tens or hundreds of amps. They can be used to replace conventional mechanical switches and relays in many d.c. and a.c. power control systems, and can readily be used to control electric lamps, motors, heaters and alarms. They have a multitude of uses in the home and in industry.

This chapter explains the basic characteristics of these outstandingly useful devices, and shows a selection of basic circuits in which they can be used for demonstration and educational purposes.

The s.c.r.: basic theory

The s.c.r., or silicon controlled rectifier, is a four-layer pnpn silicon semiconductor device, and is represented by the symbol shown in *Figure 1.1a*. Note that this symbol resembles that of a normal rectifier, but has an additional terminal known as the 'gate'. The s.c.r. can be made to act as either an open-circuit switch or as a silicon rectifier, depending on how its gate is used.

Figure 1.1b shows the transistor equivalent circuit of the s.c.r. Essentially, the circuit is that of a complementary regenerative switch, in which the collector current of npn transistor Q_1 feeds directly into the base of pnp transistor Q_2 and the collector current of Q_2 feeds into the base of Q_1. This equivalent circuit is of great value in understanding s.c.r. characteristics.

2 BASIC PRINCIPLES AND PROJECTS

Figure 1.1c shows the basic connections for using the s.c.r. as a switch in d.c. circuits. The load is connected in series with the s.c.r. anode and cathode and is connected across the supply so that the anode

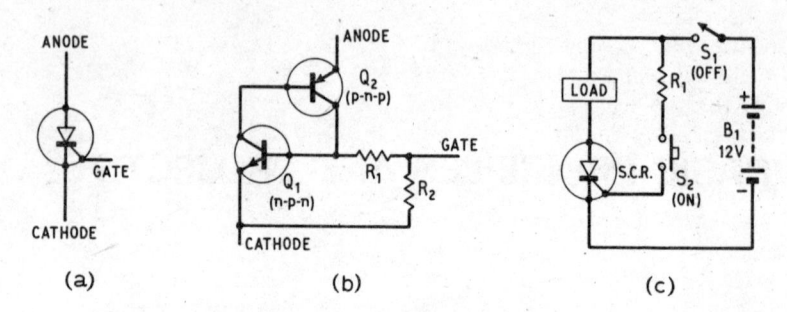

Figure 1.1a S.C.R. symbol.
Figure 1.1b Transistor equivalent circuit of the s.c.r.
Figure 1.1c Basic connections for using the s.c.r. as a d.c. switch.

is positive relative to the cathode. The gate can be connected to the positive supply line via R_1 and S_2.

The basic characteristics of the s.c.r. can be readily understood by referring to *Figures 1.1b* and *1.1c,* and are as follows.

(1). When power is first applied to the s.c.r. (by closing S_1 in *Figure 1.1c*) the s.c.r. is 'blocked', and acts (between anode and cathode) like an open-circuit switch. Looking at *Figure 1.1b* it can be seen that this action is due to the fact that Q_1 base is shorted to the cathode via R_1 and R_2, so Q_1 is cut off and passes negligible collector current into the base of Q_2, which is hence also cut off and passes negligible current into Q_1 base. Both transistors are thus cut off under this condition, and only a small leakage current flows between the anode and cathode of the device.

(2). The s.c.r. can be turned on and made to act like a normal silicon rectifier by applying a positive current to the gate by closing S_2. When this gate current is applied the s.c.r. regenerates and turns on very rapidly, and the full load current flows between the anode and cathode. As in the case of a normal silicon rectifier, a saturation potential of one or two volts is developed between the anode and cathode of the s.c.r. when it is on. Referring to *Figure 1.1b*, these characteristics can be explained as follows.

When the gate is made positive with respect to the cathode, gate current flows via R_2 and via R_1 and the base-emitter junction of Q_1. If the gate current is sufficiently large Q_1 is biased on and its collector current feeds into the base of Q_2. This base current is amplified by Q_2

and is fed back into the base of Q_1, where it is again amplified and fed to Q_2. A regenerative action thus takes place, and both transistors switch rapidly into saturation. Under this condition the anode-to-cathode saturation voltage is equal to the sum of the Q_2 saturation voltage and the Q_1 forward base-emitter voltage, and amounts to one or two volts. Modern s.c.r.s have typical turn-on times of a few microseconds.

(3). Once the s.c.r. has been turned on and is conducting in the forward direction the gate loses control, and the s.c.r. then remains on even if the gate drive is completely removed. Thus, only a brief positive gate pulse is needed to turn the s.c.r. on. Considerable current gain is available between the gate and anode of the device, so very small amounts of gate power can be used to control very high power values in the external load.

The above points should be self-evident from *Figure 1.1b*. A small drive current must be fed into the base of Q_1 via R_1 and the gate terminal to cause the circuit to regenerate initially, but once regeneration is complete the Q_1 base drive necessary to sustain the circuit in the saturated state is provided via Q_2 collector, so the circuit self-latches and remains on even when the gate drive is removed. R_1 ensures that base current continues to flow in Q_1 even if the gate-cathode terminals are shorted or reverse biased, so the gate loses all control once the circuit has switched into the self-latching mode.

Since Q_1 and Q_2 are connected as a positive feedback network they produce a combined loop gain equal to the product of their individual gains. Typically, s.c.r.s give current gains, between gate and anode, of the order of one thousand times. Milliamps of gate current can be used to control amps of anode current. The gate of the s.c.r. has non-linear characteristics similar to those of the base of a normal silicon transistor (as is self-evident from *Figure 1.1b*), and gate potentials of the order of one volt are sufficient to turn the s.c.r. on. S.c.r.s can be used to switch potentials up to several hundred volts, so high voltage gains are available from the device. Thus, extremely high power (volts x amps) gains are available from s.c.r.s, and milliwatts of gate power can be used to control kilowatts of load power.

(4). Once the s.c.r. has self-latched into the on state it can only be turned off again by momentarily reducing its anode current to zero, or below a value known as the 'minimum holding current'. Since turn-off occurs whenever the current is reduced below this critical value, it follows that turn-off occurs automatically in a.c. circuits near the zero-crossing point at the end of each half-cycle. The fact that turn-off occurs when the main current falls below a minimum holding value, rather than to zero, is due to the fact that the current gains of the two transistors in the *Figure 1.1b* circuit fall off as the anode current is

4 BASIC PRINCIPLES AND PROJECTS

reduced, and the circuit unlatches when the anode current falls to such a low value that the combined loop gain of the two transistors falls below unity. Minimum holding current values are typically of the order of a few milliamps.

(5). A certain amount of internal capacitance exists between the anode and gate of the s.c.r. Consequently, if a sharply rising voltage is applied to the s.c.r.s anode this internal capacitance can cause part of the rising anode signal to break through to the gate and thus trigger the s.c.r. on. This 'rate-effect' turn-on can be caused by supply-line transients, and sometimes occurs at the moment that supplies are switch-connected to the s.c.r. anode. Rate-effect problems can usually be overcome by wiring a simple $R-C$ smoothing network between the anode and cathode of the s.c.r., to limit the rate of rise to a safe value.

The s.c.r.: basic d.c. circuits

Having examined the basic characteristics of the s.c.r. we can now go on to look at a few basic circuits that can be used to demonstrate these characteristics in a practical way.

Figure 1.2a shows a simple d.c. on-off circuit controlling a 12 V 500 mA lamp. The lamp can be replaced with an alternative type of

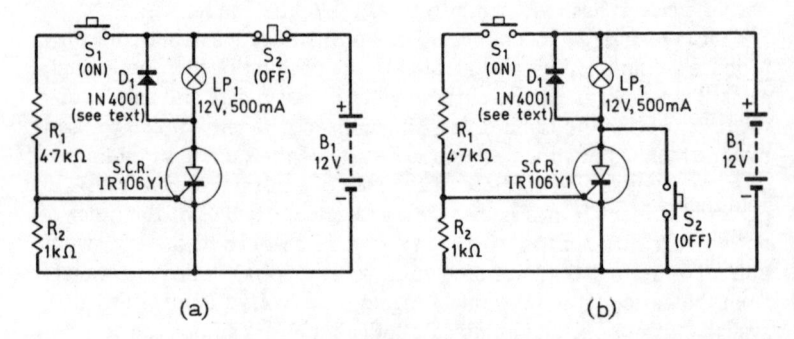

Figure 1.2a Simple d.c. on-off circuit.
Figure 1.2b Alternative d.c. on-off circuit.

load if required, but if an inductive load is used it must be shunted by a damping diode (D_1), so that the circuit is not damaged by back e.m.f.s as the s.c.r. is switched on or off. The s.c.r. used here can handle anode currents up to 2 A, and can be turned on by gate currents as low as a few hundred microamps. Gate current is provided via limiting resistor R_1 and push-button switch S_1. R_2 is wired between the s.c.r.s gate and cathode to improve circuit stability, in the same way as a resistor may

BASIC PRINCIPLES AND PROJECTS 5

be wired between the base and emitter of a conventional transistor for the same purpose.

The *Figure 1.2a* circuit can be turned on by briefly closing S_1. Once the circuit has been turned on it self-latches and stays on when S_1 is opened again. The circuit can be turned off by momentarily reducing the anode current to zero by opening S_2. Alternatively, the circuit can be turned off by momentarily applying a short between the s.c.r.s anode and cathode, as shown in *Figure 1.2b*.

Figure 1.3 shows a third way of achieving s.c.r. turn-off. Here, once the s.c.r. has been turned on, C_1 charges up via R_3 and the s.c.r. to

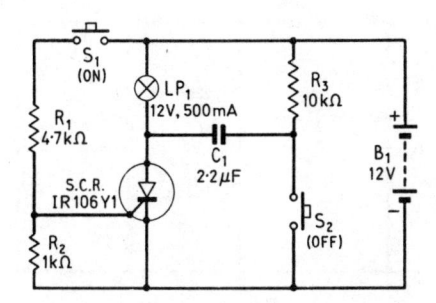

Figure 1.3. Capacitor-turn-off circuit.

almost the full supply line potential, the R_3 end going positive. When S_2 is operated the positive end of C_1 is clamped to ground, and the capacitor charge forces the s.c.r. anode to momentarily swing negative, thereby reverse-biasing the s.c.r. and causing it to turn off. The capacitor charge leaks away rapidly under this condition, but has to hold the s.c.r. anode negative for only a few microseconds to ensure complete turn-off. Note that if S_1 is held down after the charge has leaked away, the capacitor then starts to charge in the reverse direction

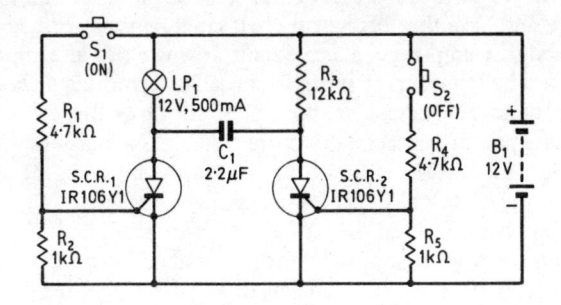

Figure 1.14. Capacitor-turn-off circuit with s.c.r. slaving.

6 BASIC PRINCIPLES AND PROJECTS

via the lamp load: C_1 must thus be a non-polarised type, such as Mylar or polyester.

A variation of the capacitor turn-off circuit is shown in *Figure 1.4*. Here a slave s.c.r. is used to replace S_2 of *Figure 1.3*, and capacitive turn-off of SCR_1 is achieved by briefly driving SCR_2 on via a low-current pulse from S_2. SCR_2 turns off once S_2 is released, since the anode current provided by R_3 is lower than the SCR_2 holding current.

Figure 1.5 shows the circuit of an s.c.r. bistable or flip-flop driving two independent lamp loads. Assume that SCR_1 is on and SCR_2 is off,

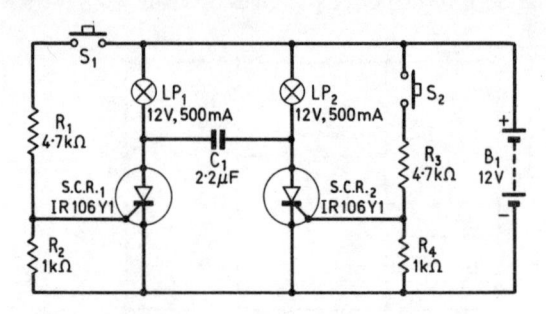

Figure 1.5. S.C.R. bistable or flip-flop.

so that C_1 (non-polarised) is fully charged with its LP_2 end positive. The state of the circuit can be changed by momentarily operating S_2. SCR_2 is then driven on via its gate, and as it goes on it drives SCR_1 off capacitively via its anode. C_1 then recharges in the reverse direction. Once recharging is complete the state of the circuit can again be changed by briefly operating S_1, thus driving SCR_1 on via its gate and driving SCR_2 off capacitively via its anode. The flip-flop process can be repeated *ad infinitum*.

All the d.c. circuits that we have looked at so far have used simple fixed loads, and have thus been of the self-latching type. *Figure 1.6a*, however, shows a simple d.c. alarm circuit driving a self-interrupting load such as a bell, buzzer, or siren. When self-interrupting devices such as these are connected across a supply a current flows through a built-in solenoid via a pair of contacts; this current induces a magnetic field in the solenoid, and causes a striker to fly outwards and open the contacts, causing the current to fall to zero and making the magnetic field collapse. Once the field has collapsed the striker falls back again and the contacts close, so current is again applied to the solenoid and the action repeats.

Consequently, this type of load acts like a switch that repetitively opens and closes rather rapidly. When such loads are connected in the *Figure 1.6a* circuit, therefore, the circuit does not self-latch in the normal way, and the alarm operates only so long as S_1 is closed. Due to the inductive nature of such loads a damping diode must be wired across them when they are used in s.c.r. circuits, as shown in the diagram.

The alarm circuits can be made to self-latch, if required, by simply wiring a 470-ohm resistor in parallel with the alarm, as shown in

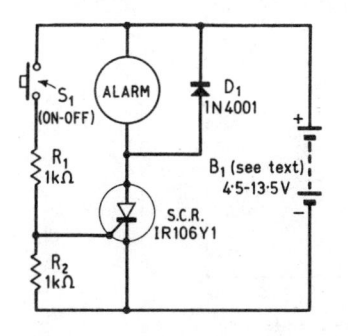

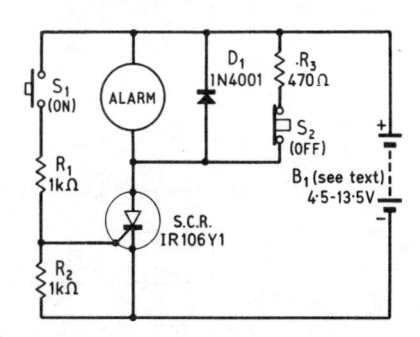

Figure 1.6a. Simple non-latching alarm circuit.

Figure 1.6b. Simple self-latching alarm circuit.

Figure 1.6b. In this case the anode current of the s.c.r. does not fall to zero when the alarm self-interrupts, but falls to a value dictated by R_3 and the battery voltage. If this current is in excess of the s.c.r.s holding current the s.c.r. self-latches. The circuit can be unlatched by briefly operating S_2, so that the anode current falls to zero when the alarm enters a self-interrupting stage, and the s.c.r. turns off.

The actual alarm used in the *Figure 1.6* circuits can be any low-voltage (3 to 12 V) self-interrupting bell, buzzer, or siren that draws an operating current of less than 2 A. The supply battery should give a voltage roughly 1·5 V greater than the normal operating voltage of the alarm device, to compensate for the saturation voltage that is 'lost' across the s.c.r. when it is on.

Figure 1.7 shows a circuit that can be used to demonstrate the rate-effect turn-on of the s.c.r., and a method that can be used for rate-effect suppression. Here, the s.c.r. uses a 3 V lamp as its anode load, and is connected across the 4·5 V battery supply via S_1. A 4·5 V domestic door bell can be connected across the supply via S_2, and enables transient modulation to be applied to the supply line and thus to the anode of the s.c.r. This modulation can cause rate-effect turn-on of the IR106Y1 s.c.r., which has a critical rate-of-rise value of 20 V/μs.

To demonstrate the rate-effect, open S_3 so that the suppressor network is removed from the s.c.r., connect the supply by closing S_1, and then close S_2 so that the bell rings. As the bell operates it applies

8 BASIC PRINCIPLES AND PROJECTS

high transient modulation to the positive supply line, and this modulation appears at the s.c.r. anode and should trigger the s.c.r. and the lamp on. If the lamp does not go on, it may be that the battery has a low internal resistance. If so, try connecting a 0·5-ohm resistor in series with

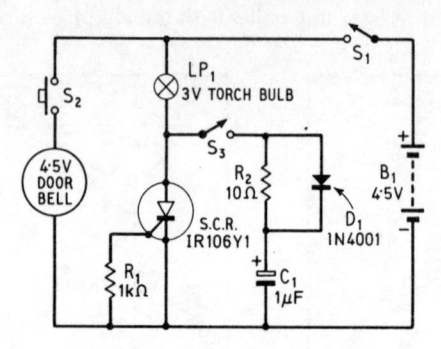

Figure 1.7. Rate-effect demonstration circuit.

the battery, so that the correct operation is obtained. Once the s.c.r. and lamp have been triggered on by rate-effect they can be turned off again by briefly opening S_1.

Once the turn-on rate-effect has been demonstrated the effects of the suppressor network can be shown by closing S_3 and operating the bell again via S_2. When S_3 is closed the lamp resistance and C_1 act as a smoothing network that reduces the rate-of-rise of the anode modulation signal to a safe value, and the s.c.r. does not go on when the bell operates. If the s.c.r. still goes on under this condition, increase the C_1 value until the correct results are obtained. C_1 charges to the full supply line potential when S_1 is closed, and R_2 is wired in series with C_1 to limit its discharge current into the s.c.r. anode to a safe value. D_1 makes R_2 appear as a virtual short-circuit to sharply rising anode signals, however, so R_2 does not effect the time constant of the R–C rate-effect suppression network formed by C_1 and the lamp resistance. In many applications D_1 can be eliminated from the suppressor circuit.

The s.c.r.: basic a.c. circuits

Figure 1.8 shows a basic half-wave a.c. on-off circuit driving a 100 W lamp from a 120 V or 240 V a.c. power line. With S_1 open zero drive is applied to the s.c.r. gate, so the s.c.r. and lamp are off. Suppose, however, that S_1 is closed. At the start of each positive half-cycle the s.c.r. is off, so the full available line voltage is applied to the gate via the lamp and D_1 and R_1. Shortly after the start of the half-cycle sufficient voltage is

BASIC PRINCIPLES AND PROJECTS 9

available to trigger the s.c.r., and the s.c.r. and lamp go on. As the s.c.r goes on its anode voltage falls to near zero, thus removing the gate drive current. Since a substantial anode current is flowing in the s.c.r. at this time, however, the s.c.r. remains fully latched on for the duration of the half-cycle. The s.c.r. automatically turns off when the half-cycle ends and the anode current falls to zero.

This process repeats, with the s.c.r. triggering on shortly after the start of each positive half-cycle, so long as S_1 is closed, and half-wave

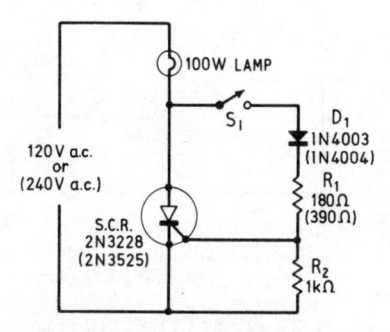

Figure 1.8. Line-driven half-wave on-off circuit.

power is applied to the lamp. The s.c.r. and lamp turn off when S_1 is re-opened, since s.c.r. turn-off occurs automatically at the end of each positive half-cycle.

Diode D_1 in this circuit prevents reverse bias being applied to the s.c.r. gate on negative half-cycles of line voltage. R_1 is given a low value so that the s.c.r. can be turned on as early as possible in each positive half-cycle, but is large enough to limit the peak gate current to a safe value in the event of S_1 first being closed when the instantaneous line voltage is at its peak value. Note that although very high peak voltage and and current values may be applied to R_1, they are applied for only the few microseconds that the s.c.r. takes to turn on; the mean power dissipation of R_1 is thus quite low, and this component can safely be given a half-watt rating.

S.C.R.s can be used in a variety of ways to give full-wave a.c. control. In *Figures 1.9* and *1.10* the a.c. is converted to rough (unsmoothed) d.c. via the bridge rectifier D_1-D_4, and the rough d.c. is applied to the s.c.r. With S_1 open the s.c.r. is off, so zero current flows through the bridge and the load. When S_1 is closed the s.c.r is driven on shortly after the start of each half-cycle of rough d.c., and full-wave power is applied to the load. As the s.c.r. goes on in each half-cycle the gate drive is automatically removed, but the s.c.r. stays latched on for the duration of each half-cycle, as already described. The s.c.r. switches off

10 BASIC PRINCIPLES AND PROJECTS

automatically at the end of each half-cycle as its anode current falls to zero, so power is automatically removed from the load when S_1 is opened.

Note in the *Figure 1.9* circuit that the load is connected in the d.c. side of the bridge, and the circuit is thus used to drive a d.c. load. A fuse

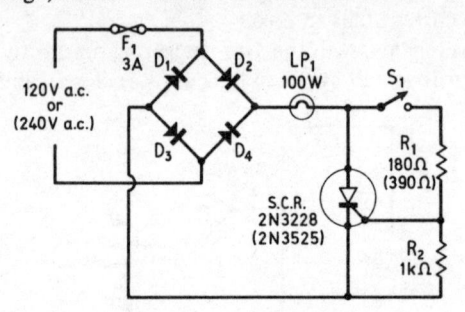

Figure 1.9. Full-wave on-off circuit with a d.c. load.
$D_1 - D_4$ = 3A 200 p.i.v. silicon rectifiers.
(3A 400 p.i.v. silicon rectifiers.)

is placed in the a.c. side of the bridge to give protection in the event of a failure of one or more of the bridge components. In the *Figure 1.10* circuit the load is placed in the a.c. side of the bridge, so the circuit is used to control an a.c. load. A fuse is not required in this case, since the load itself limits current to a safe value in the event of a component failure.

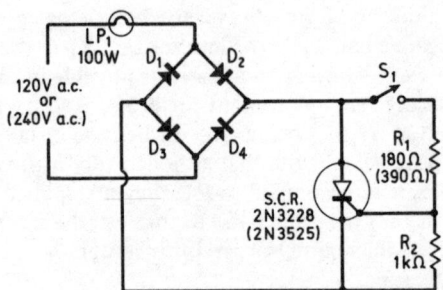

Figure 1.10. Full-wave on-off circuit with a.c. load.
$D_1 - D_4$ = 3A 200 p.i.v. silicon rectifiers.
(3A 400 p.i.v. silicon rectifiers.)

Finally, *Figure 1.11* shows how two s.c.r.s can be connected in inverse parallel to give direct full-wave switching of the load. With S_1 open zero gate drive is applied to either s.c.r., so zero power is applied to the load. When S_1 is closed gate current is applied to SCR_1 on positive half-cycles via D_1 and R_2, and SCR_1 is driven on. On negative

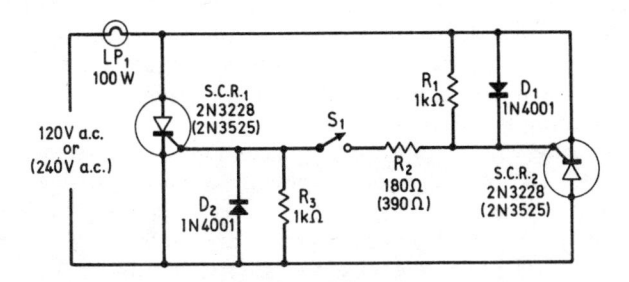

Figure 1.11. Full-wave line switch using two s.c.r.s.

half-cycles gate drive is applied to SCR_2 via D_2 and R_2, and SCR_2 is driven on. Full-wave control is thus obtained.

The triac: basic principles and projects

S.C.R.s are unidirectional thyristor devices. They pass current in one direction only, from anode to cathode. Triacs are also members of the thyristor family, but are bidirectional devices. They can pass current in either direction. For most practical purposes a triac can be regarded as two conventional s.c.r.s connected in inverse parallel within a single three-terminal package, but so arranged that the two s.c.r.s share a single gate terminal. The triac can be used as a solid-state power switch that is normally off but can be turned on via a suitable gate signal.

Figure 1.12a shows the symbol that is used to represent the triac, and *Figure 1.12b* shows a basic connection for using the device as an a.c. power switch. The load is wired in series with the triacs main

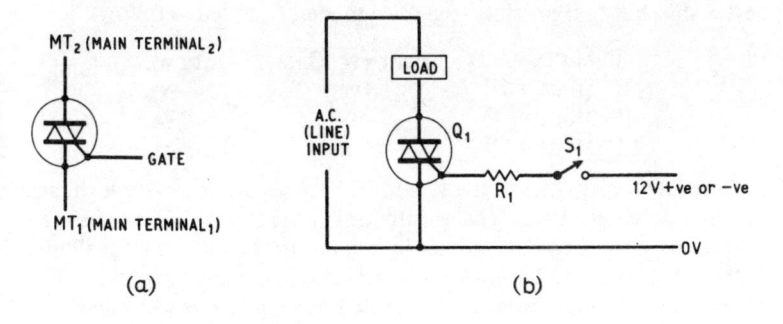

Figure 1.12a. Triac symbol.
Figure 1.12b. Basic triac switching circuit with d.c. gate drive.

terminals, and the series combination is wired directly across the a.c. power line: D.C. gate drive can be applied to the triac by closing S_1.

The triac exhibits characteristics very similar to those of a pair of s.c.r.s connected in inverse parallel. Referring to *Figure 1.12b*, these characteristics are as follows.

(1). Normally, with no gate signal applied, the triac is off and acts (between MT_1 and MT_2) like an open-circuit switch. It passes negligible current in either direction, and zero power is applied to the load.

(2). If MT_2 is appreciably positive or negative relative to MT_1 the triac can be turned on (so it acts like a closed switch between MT_1 and MT_2) by applying a suitable trigger or bias signal to its gate via S_1. The device takes only a few microseconds to switch fully on.

(3). Once the triac has been turned on via its gate it self-latches, and remains on so long as main-terminal currents continue to flow. Only a brief pulse of gate current is thus needed to turn the triac on.

(4). Once the triac has self-latched the gate loses control and the device can only be turned off again by momentarily reducing its main-terminal currents below a near-zero minimum holding value. When the triac is used as an a.c. power switch, therefore, turn-off occurs automatically at the zero-crossing point at the end of each half-cycle as the main-terminal currents fall to zero.

(5). A saturation potential of one or two volts is developed across the triacs main-terminals when the device is switched into the on state. Thus if the triac is used to switch a 10 A load from a 240 V r.m.s. supply, and the triac has a saturation potential of 1·5 V, approximately 2 400 W can be switched by the triac, but only 15 W are 'lost' in the triac itself: This 15 W can be readily dissipated by connecting the triac to a suitable heat sink. The triac thus makes an efficient a.c. power switch.

(6). The triac can be turned on by either a positive or a negative gate signal, irrespective of the polarities of the main-terminal voltages. The device thus has four possible triggering modes, signified as follows:

I^+ Mode = MT_2 current +ve, Gate current +ve.
I^- Mode =   "   "   +ve,   "   "   −ve.
III^+ Mode = "   "   −ve,   "   "   +ve.
III^- Mode = "   "   −ve,   "   "   −ve.

The gate sensitivities in the I^+ and III^- modes are relatively high, and are approximately equal. The sensitivities in the I^- and III^+ modes are also roughly equal to one another in modern triacs, but are only about half as great as in the I^+ and III^- modes. The gate exhibits non-linear characteristics similar to those of the s.c.r. Gate potentials of one or two volts and currents of a few tens of milliamps are sufficient to turn on a triac and control several amps of load current. Considerable power gain is thus available between the triac gate and the load.

BASIC PRINCIPLES AND PROJECTS 13

Having examined the basic characteristics of the triac we can now go on to look at a few basic switching circuits that can be used to demonstrate these characteristics in a practical way.

Figure 1.13 shows the practical circuit of a simple d.c. triggered triac power switch. When S_1 is open zero current flows to the gate of the triac, which is thus off. Zero power is applied to the load under this condition. When S_1 is closed gate drive is applied to the triac via R_1 and the battery, and the triac is driven on and acts like a closed switch. Full power is thus applied to the load under this condition.

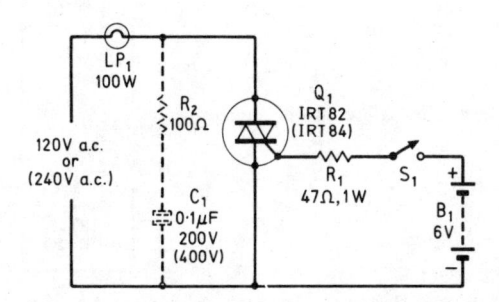

Figure 1.13. D.C.-triggered triac line switch.

Note in this circuit that positive gate drive is applied to the triac whenever S_1 is closed, irrespective of the MT_2 polarity. The triac is thus gated in the I^+ and III^+ modes alternately. The R_1 value is made low enough to trigger the triac even in the relatively insensitive III^+ mode.

The lamp load of the *Figure 1.13* circuit can, if desired, be replaced with an alternative type of load. If an inductive load such as a motor is used, the R_2-C_1 network must be wired in place as indicated. Alternating currents and voltages are inherently out of phase with one another in all inductive loads: When the triac turns off as its main-terminal current falls to zero in each half-cycle, therefore, these phase differences cause a high value of instantaneous line voltage to be suddenly applied to the triac: If the rate-of-rise of this 'commutation' or turn-off voltage exceeds a critical value it can trigger the triac back on again by rate-effect, so that the triac effectively stays permanently locked on. This problem can be overcome by wiring the $R_2 - C_1$ rate-effect suppression or 'snubber' network across the triac to limit its rate-of-rise to a safe value of about $1\,\text{V}/\mu\text{s}$.

Figure 1.14 shows how the triac can be used as a simple line switch with line-derived triggering. When S_1 is open zero drive is applied to the triac gate, so the triac and lamp are off. Suppose, however, that S_1 is closed. At the start of each half-cycle the triac is off, so the full line voltage is applied to the gate via the lamp and R_1. Shortly after the start

14 BASIC PRINCIPLES AND PROJECTS

of the half-cycle sufficient drive is available to trigger the triac, and the triac and lamp go on. As the triac goes on it saturates, thus removing the gate drive. Once it has turned on the triac self-latches and stays on until the end of the half-cycle, when it automatically turns off again as its main-terminal current falls to zero.

This process repeats, with the triac triggering on shortly after the start of each half-cycle, so long as S_1 is closed, and full power is applied to the lamp. The triac and lamp turn off again when S_1 is opened. Note

Figure 1.14. Line-triggered line switch.

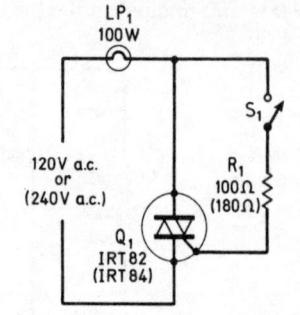

in this circuit that the triggering occurs alternately in the sensitive I^+ and III^- modes. R_1 is given a low value so that the triac is turned on as early as possible in each half-cycle, but the value is large enough to limit the peak gate current to a safe value if S_1 is first closed at the moment when the instantaneous line voltage is at its peak value.

Finally, *Figure 1.15* shows how the triac can be wired as a simple three-way line switch, giving *Off–half-wave–On* operation. When S_1 is in the '1' position zero drive is applied to the triac gate, so the triac

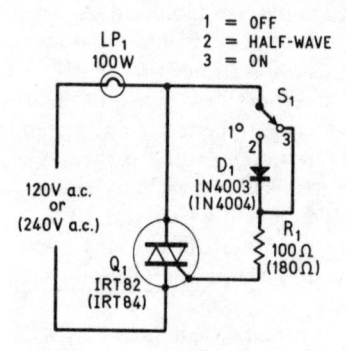

Figure 1.15. Three-way line switch.

and lamp are off. In position '2' gate drive is applied via D_1 on positive half-cycles only, so the lamp is operated at half-wave power. In position '3' gate drive is applied on all half-cycles, and the circuit operates in the same way as *Figure 1.14*. Full power is applied to the lamp load.

Phase-triggered power-control principles

The s.c.r. and triac circuits that we have looked at so far have all been used to give a simple on-off type of power control in which either full or zero power is applied to the load. S.C.R.s and triacs are, however, readily capable of giving fully variable control of a.c. loads, and can be used to infinitely and smoothly vary the load power all the way from zero to maximum. They are thus suitable for use in applications such as lamp dimmers, electric motor speed controllers, and electric heater power controllers.

The most widely used system of s.c.r. and triac variable power control in a.c. circuits is known as the 'phase triggering' system. The principle is illustrated in *Figure 1.16*.

Figure 1.16a shows the basic phase-triggered circuit, using a triac as the power control element. The load is wired in series with the triac, and the combination is connected across the a.c. power line. The triac gate-trigger signal is derived from MT_2 via a variable phase-delay network and

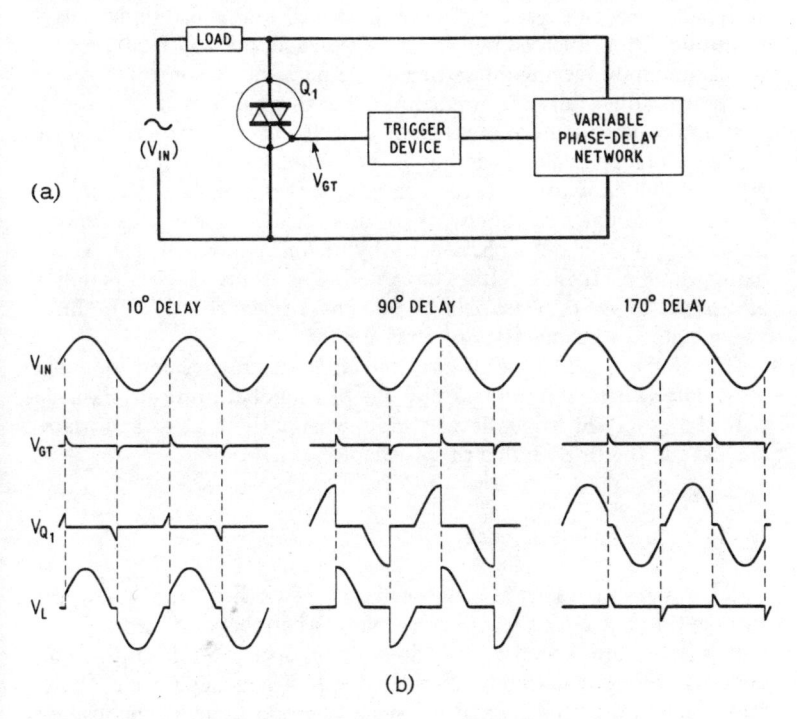

Figure 1.16a. Basic phase-triggered variable power-control circuit.
Figure 1.16b. Waveforms of the *Figure 1.16a* circuit at various phase-delay settings.

a trigger device. The phase-delay network enables the a.c. voltage to the input of the trigger device to be delayed relative to that on MT_2 by an amount fully variable from (ideally) 0° to 180°, i.e., by as much as one half-cycle of line voltage.

The trigger device is a voltage-operated 'switch' that triggers on and fires the triac when a preset voltage is reached at the output of the phase-delay network, or at the end of the pre-set phase-delay period.

Figure 1.16b shows the waveforms that occur in different parts of the circuit at three different settings of phase delay. Thus, if the phase delay network is set for only a 10° delay the triac is triggered on 10° after the start of each half-cycle, and then self-latches and stays on for the remaining 170° of each half-cycle. Almost the full available line power is thus applied to the load under this condition.

If the circuit is set for a phase-delay of 90°, the triac does not turn on until half-way through each half-cycle, and only half of the maximum possible power is applied to the load. Finally, if the circuit is set for a 170° delay the triac does not turn on until 10° before the end of each half-cycle, and very little power is applied to the load under this condition. Thus, the load power can be varied all the way from zero to maximum by varying the setting of the phase-delay control. Since the triac is either fully off or is saturated at all times very little power is 'lost' in the device, and very efficient variable power control is obtained.

The actual phase-delay section of the *Figure 1.16* circuit can take either of two basic forms. It can consist of either a single or multiple $R-C$ variable phase-shift network, or of an $R-C$ variable time-delay network that simulates a phase-delay by its time-equivalent: e.g., a half-cycle of 50 Hz line voltage has a period of 10 ms, giving a period of 55·5 μs per degree of phase delay; a 90° phase-triggered delay can thus be simulated by a time delay of 5 ms, etc.

The actual trigger device used in the circuit is usually some kind of solid-state switch, and can take any one of a number of forms. Basic details of a variety of trigger devices and special-purpose s.c.r.s and triacs are given in the final section of this chapter.

Radio-frequency interference

S.C.R.s and triacs act as high-speed power switches. They have typical turn-on times of only a few microseconds. When they are used to switch power into a load these high switching speeds result in the generation of a series of harmonically related radio-frequency signals. The magnitude of the fundamental r.f. signal is proportional to the magnitude of the device switching current, and may be so great that it causes interference on a.m. radios.

BASIC PRINCIPLES AND PROJECTS 17

Two basic types of radio-frequency interference (r.f.i.) may be generated by s.c.r.s and triacs when they are used in line-powered switching applications. One of these is radiated r.f.i. which is radiated directly into the air as a radio signal. In most cases this type of r.f.i. is of such low intensity that it does not cause significant interference with a.m. radios unless they are placed very close to the source of radiation. If radiated r.f.i. is troublesome, it can be minimised by mounting the thyristor circuitry in a screened container.

The second and more troublesome type of radio frequency interference is conducted r.f.i., which is carried through the power lines and may affect radio and TV sets connected to the same power lines. Trouble from this type of r.f.i. can be minimised by connecting a simple $L-C$ filter in series with the power line, so that the conducted high-frequency harmonics of the basic switching signal are reduced to insignificant levels.

Looking back at some of the triac switching circuits described earlier, the following points should be noted concerning r.f.i.

In the *Figure 1.13* circuit the triac is off when S_1 is open, so no r.f.i. is generated. When S_1 is closed the triac is permanently on, and does not perform a high-speed switching function, so zero r.f.i. is again generated. The triac does switch on rapidly at the instant that S_1 is first closed, however, and thus generates a brief pulse of r.f.i. at that instant. The magnitude of this pulse is proportional to the magnitude of the instantaneous turn-on current, and thus line voltage, at the moment that S_1 is closed, and may be of substantial proportions. Since the pulse is very brief and only occurs as S_1 is operated, however, the resulting r.f.i. is unlikely to cause annoyance. This initial switch-on r.f.i. can only be completely eliminated by using a synchronous zero-voltage gating technique, in which gate drive can only be applied to the triac in the brief periods when the line voltage is at or near the zero-voltage cross-over point near the start of each half-cycle.

The *Figure 1.14* circuit also generates a brief pulse of r.f.i. as S_1 is closed, in the same way as described above. In addition, however, the circuit generates a continuous r.f.i. signal of low intensity while S_1 is closed. This r.f.i. is caused by the fact that the triac does not turn on in each half-cycle until the line voltage has risen to a value (of a few volts) sufficient to trigger the device on via R_1, at which point the triac switches sharply into saturation. A switching pulse is thus generated shortly after the start of each half-cycle of line voltage so long as S_1 is closed. This pulse is, however, of low intensity; and will not effect a.m. radios unless they are placed very close to the triac circuit.

Finally, consider the case of the phase-triggered variable power control circuit of *Figure 1.16*. This circuit also triggers on some time after the start of each half-cycle, and thus generates continuous r.f.i. In this case, however, very high switching currents may be involved, particularly when the circuit is set for a 90° delay, and high levels of r.f.i. may thus be generated. This r.f.i. is predominantly of the conducted type, and must be suppressed if it is not to interfere with a.m. radios. If the circuit is used with an inductive load such as a motor, the inductance itself will act as an effective r.f.i. suppressor. If a resistive load such as a lamp is used, the circuit must be fitted with a simple $L-C$ suppressor to filter out the higher harmonic signals.

Miscellaneous thyristors and trigger devices

S.C.R.s and triacs are two members of a wide range of solid-state devices in the thyristor family. In this final section of this chapter we shall take a brief look at the characteristics of six other members of the family, and at two special triggering devices that are associated with them, as follows.

The l.a.s.c.r. This device is simply a Light Activated s.c.r. All semiconductor junctions are sensitive to light, and most active semiconductor devices are consequently shrouded in an opaque material to exclude unwanted light effects. The l.a.s.c.r., however, is deliberately manufactured so that one or more of its junctions can be readily exposed to light, so that the s.c.r. can be directly triggered on by a light source.

The s.c.s. The s.c.s., or Silicon Controlled Switch, uses the symbol shown in *Figure 1.17a*. Note that this symbol resembles that of a normal s.c.r., but has an additional gate terminal near its anode. *Figure 1.17b* shows the approximate equivalent circuit of the device. Note that this equivalent circuit is very similar to that of the s.c.r. shown in *Figure 1.1b*, except that the base of Q_2 is available externally at the anode gate terminal. The characteristics of the device are in fact very similar to those of the s.c.r., with the following exceptions:

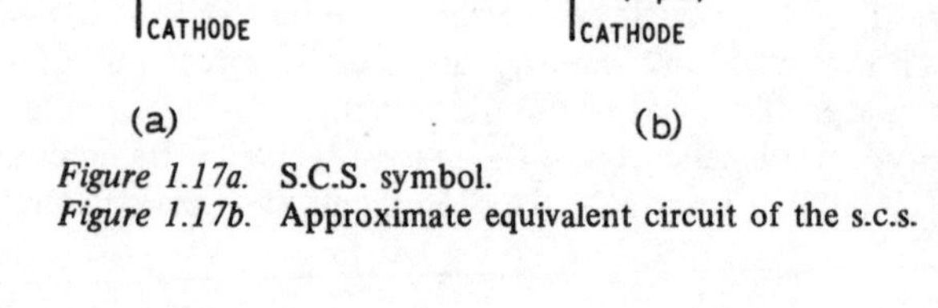

Figure 1.17a. S.C.S. symbol.
Figure 1.17b. Approximate equivalent circuit of the s.c.s.

BASIC PRINCIPLES AND PROJECTS 19

(a) The device can be turned on by forward biasing the base-emitter junction of either Q_1 or Q_2 via the external gate terminals, i.e., the device can be turned on by applying a positive pulse to the cathode gate or by applying a negative pulse to the anode gate.

(b) Once the device has turned on it can be turned off again by reverse biasing the base-emitter junction of either Q_1 or Q_2, i.e., by applying a negative pulse to the cathode gate or a positive pulse to the anode gate.

The l.a.s.c.s. This device is simply a Light Activated s.c.s., i.e., it is an s.c.s. mounted in an encapsulation that enables light to reach one or more of its junctions, so that the s.c.s. can be triggered directly by light.

The s.u.s. This device is a Silicon Unilateral Switch, and uses the symbol shown in *Figure 1.18a. Figures 1.18b* and *1.18c* show two alternative equivalent circuits of the device, which is essentially an s.c.r. with

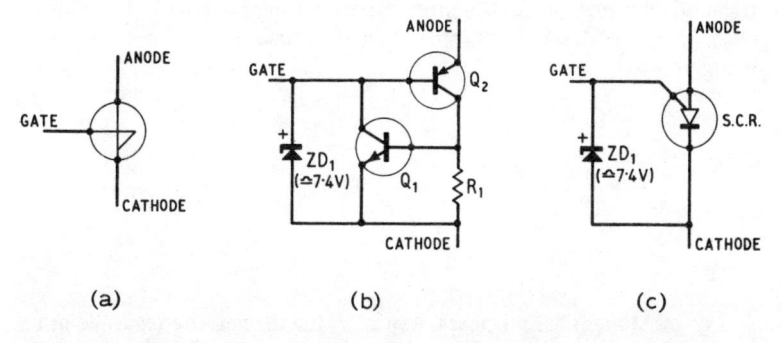

Figure 1.18a. S.U.S. symbol.   *Figure 1.18b and c.* S.U.S. equivalent circuits.

an anode gate (instead of a cathode gate) and with a built-in zener diode between the gate and cathode. Normally the device is used with the gate connection left open. In this case the device acts as a voltage-triggered switch which turns on and saturates as soon as the anode voltage becomes sufficiently positive to cause the internal zener diode to start to break down via the base emitter junction of Q_2. The typical turn-on voltage of the device is 8 V. The turn-on voltage can be reduced by wiring a lower voltage zener diode between the gate and cathode.

The s.b.s. The s.b.s., or Silicon Bilateral Switch, uses the symbol shown in *Figure 1.19a*. The device consists of two identical s.u.s. structures connected in inverse parallel within a single housing, as indicated by the equivalent circuit of *Figure 1.19b*. The device acts as a voltage-triggered switch that can be fired by either polarity of voltage.

20 BASIC PRINCIPLES AND PROJECTS

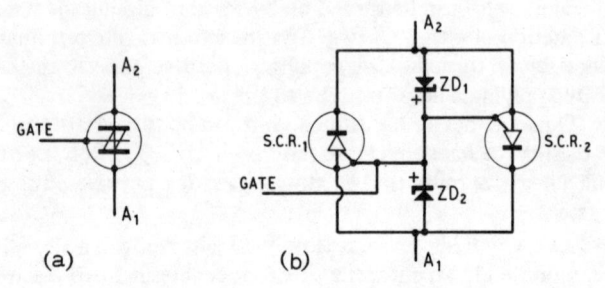

Figure 1.19a. S.B.S. symbol.   *Figure 1.19b.* S.B.S. equivalent circuit.

The diac. The diac is also a bilateral trigger device, and uses the symbol shown in *Figure 1.20*. This symbol in fact gives a slightly misleading impression of the device structure and characteristics, since the trigger diac is not a true thyristor, and uses a special three-layer transistor structure that exhibits a negative resistance characteristic at the firing points.

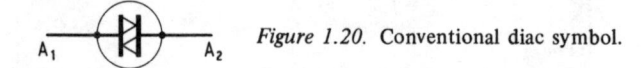

Figure 1.20. Conventional diac symbol.

For most practical purposes, however, the diac can be regarded as an s.b.s. with a high turn-on voltage (about 35 V) and a high saturation voltage (about 30 V). Thus, if a rising voltage is applied to the device via a limiting resistor, the diac acts as an open switch until the applied potential reaches 35 V, at which point the device triggers and develops a 5 V pulse across the load, the remaining 30 V being developed across the diac. The diac turns off when its current falls below a minimum holding value. The device can be connected into circuit in either polarity, and is widely used as a trigger device in phase-triggered triac variable power control applications.

Figure 1.21. Quadrac symbol.

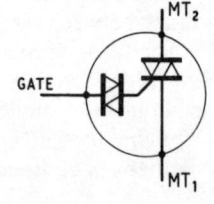

The quadrac. The quadrac is simply a triac with a diac built into its gate network, the two devices being housed in a single package. The quadrac is specifically designed for use in phase-triggered variable power control applications, and uses the symbol shown in *Figure 1.21*.

The u.j.t. The u.j.t., or Unijunction Transistor, uses the symbol shown in *Figure 1.22a*. The u.j.t. is a two-layer solid-state trigger device, and is not a true member of the thyristor family. Its characteristics can, however, be simulated by a thyristor equivalent circuit for most practical purposes, as shown in *Figures 1.22b* and *1.22c*. It can be seen that the device resembles an s.c.r. with anode gate triggering provided via a fixed potential divider formed by R_1 and R_2.

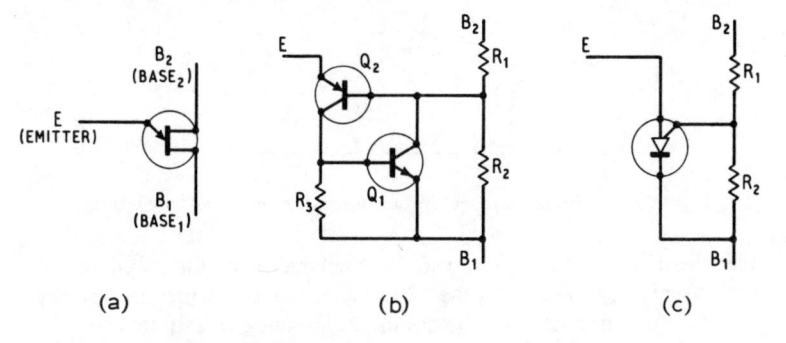

Figure 1.22a. U.J.T. symbol.
Figure 1.22b. Transistor equivalent of the u.j.t.
Figure 1.22c. Alternative equivalent of the u.j.t.

The device is normally wired in circuit so that B_2 is at a fixed positive voltage, and a variable input voltage is applied to E: Due to the voltage divider action of R_1 and R_2 a fixed 'reference' voltage is applied to Q_2 base under this condition. If the input voltage at E is below the reference voltage, therefore, Q_2s base-emitter junction is reverse biased, and both transistors are cut off. E thus appears as a high impedance to the input signal. If, on the other hand, the voltage at E rises significantly above the reference value set by R_1 and R_2, the Q_2 base-emitter junction becomes forward biased and the circuit regenerates and switches sharply on. E thus appears as a low impedance under this condition. The voltage at which turn-on occurs is known as the 'peak-point' voltage.

Figure 1.23 shows how the u.j.t. can be wired as a simple relaxation oscillator. When power is first applied to the circuit C_1 is fully discharged and the u.j.t. is off. As soon as power is applied C_1 starts to charge exponentially via R_1. Eventually, after a pre-set delay, the C_1 voltage rises to the peak-point or firing voltage of the u.j.t, which then fires and discharges C_1 rapidly into R_3 as E goes into a low-impedance saturated state. As the C_1 charge decreases a point is reached where the

22 BASIC PRINCIPLES AND PROJECTS

E current of the u.j.t. falls below the devices minimum holding or 'valley' current, and the device turns off. C_1 then starts to charge again via R_1, and the process repeats. The circuit thus acts as a free-running oscillator that generates a sawtooth waveform across C_1 and a pulse waveform across R_3. R_2 is used to enhance the circuits thermal stability.

A very important point about the *Figure 1.23* circuit concerns the selection of the R_1 value. This value must be large enough to ensure

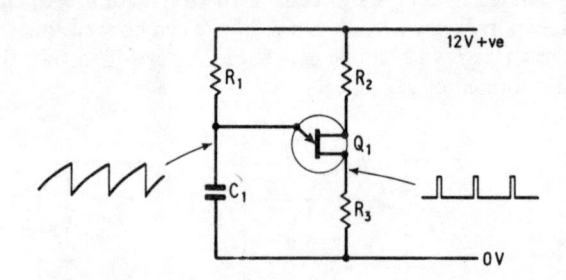

Figure 1.23. Simple u.j.t. relaxation oscillator circuit and waveforms.

that the u.j.t. can turn off once C_1 is discharged, i.e., the maximum R_1 current must be lower than the u.j.t.s valley current, which is typically about 5 mA. In most cases the minimum R_1 value is restricted to a few thousand ohms. The R_1 value must, on the other hand, be low enough to provide the u.j.t. with adequate turn-on or 'peak-point' triggering current once C_1 reaches the peak-point voltage of the u.j.t. Worst-case peak-point currents are normally of the order of a few microamps, so the maximum R_1 value is usually restricted to a few hundred thousand ohms. Note, therefore, that the R_1 value can be varied over a 100 : 1 range in this simple circuit, and that the operating frequency or periods of the oscillator can thus be varied over the same range by simply making R_1 a variable control.

CHAPTER 2

15 A.C. POWER-SWITCHING PROJECTS

Triacs are high-speed solid-state a.c. power switches. They have no moving parts to arc or wear out, and they give a high power gain between their gate-control and main terminals. They can thus be used with advantage to replace conventional mechanical switches and relays in many power-control applications, and can be made to perform a variety of sophisticated switching tasks.

Fifteen a.c. power-switching projects are described in this chapter. They include simple power switches, a self-latching switch, water or steam-operated line switches, synchronous switches, and automatic overload switches. All these projects are designed around the IRT82 or IRT84 triacs manufactured by International Rectifier. The IRT82 is a 200 V 8 A device, and can handle loads up to 960 W on 120 V a.c. lines in all the projects described here: The IRT84 is a 400 V 10 A device, and can handle loads up to 2 400 W on 240 V lines. The triacs must, of course, be fitted to suitable heat sinks when operated at high power levels. Typically, the specified triacs dissipate about 18 W when operating at full load power.

Simple power-switching projects

Figure 2.1 shows the circuit of a simple line-triggered a.c. power switch. The operation of this circuit has already been described in detail in Chapter 1, and is such that the triac turns on shortly after the start of each half-cycle via R_1 when S_1 is closed. The R_1 gate-drive is removed automatically when the triac turns on, so the mean current through S_1 is of the order of only a few milliamps, although a peak-surge current up to two amps may flow through S_1 at the instant that the switch is first closed.

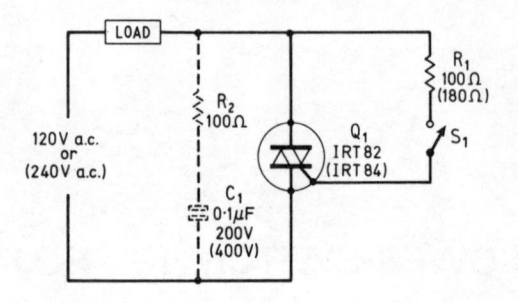

Figure 2.1. Simple a.c. power switch, line triggered.

The peak-surge turn-on current of S_1 can be reduced to less than 100 mA by using the circuit shown in *Figure 2.2*. Here the triac's gate drive is derived from $R_1 - C_1$ and silicon bilateral switch (s.b.s.). The circuit action is such that C_1 starts to charge up via R_1 at the start of each half cycle, the C_1 voltage following that of the a.c. power line, until C_1 reaches the firing voltage of the s.b.s. (about 8 V). The s.b.s. then fires and discharges C_1 into the gate of the triac via current-limiting resistor R_2, thus driving the triac on. As the triac goes on it saturates and removes all power from the gate-drive network. The triac

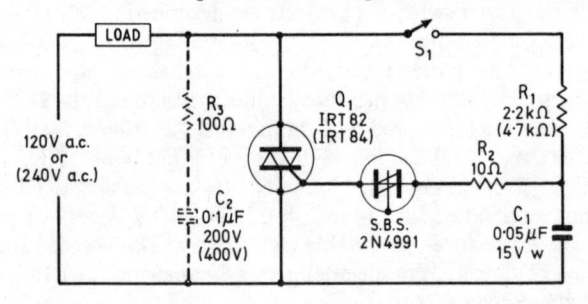

Figure 2.2. A.C. power switch with s.b.s -aided triggering.

turns off automatically at the end of the half-cycle, and the triggering process then repeats.

Note that the gate trigger current of this circuit is obtained from C_1 rather than R_1, and that R_1 can thus be given a fairly large value (to limit the peak-surge turn-on current of S_1) without causing the triac to turn on excessively late in each half-cycle.

The circuits of *Figures 2.1* and *2.2* perform as efficient power switches, but do not turn on in each half-cycle until the line voltage has risen to a value of several volts. These circuits thus generate continuous low-level r.f.i. when S_1 is closed, and this r.f.i. may interfere with a.m. radios that are placed close to the triac circuitry. This r.f.i. problem can be overcome by using d.c. triggering of the triac gate, as shown in *Figures 2.3* and *2.4*.

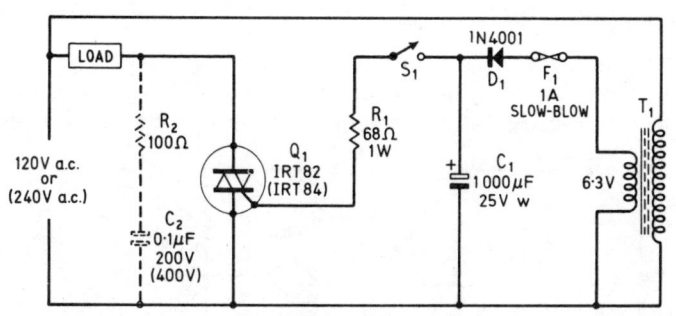

Figure 2.3. Simple a.c. power switch, d.c. triggered.

The gate drive is taken from the d.c. supply via S_1 and R_1 in the *Figure 2.3* circuit, and S_1 passes an *on* current of about 120 mA. In the *Figure 2.4* circuit the gate drive is obtained via switching transistor Q_2, which is driven on via S_1, and S_1 passes an *on* current of less than 3 mA.

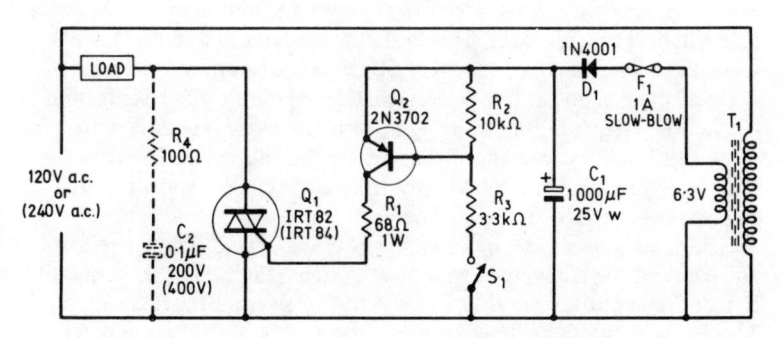

Figure 2.4. A.C. power switch with transistor-aided d.c. triggering.

The S_1 current can be further reduced by using an additional transistor stage. By replacing S_1 with suitable electronic circuitry the triac can be turned on by heat, light, sound, etc. The d.c. supply in these two circuits can be obtained from batteries, or from the a.c. power line via a suitable step-down transformer and rectifier and smoothing network, as shown in the diagrams.

26 15 A.C. POWER-SWITCHING PROJECTS

The basic circuits of *Figures 2.1* and *2.4* can be used with either resistive or inductive loads. If inductive loads are used in these circuits an $R-C$ 'snubber' network must be wired across the triacs, as shown in the diagrams.

Special power-switching projects

A disadvantage of the d.c.-triggered circuits of *Figures 2.3* and *2.4* is that they take a fairly heavy current (roughly 120 mA) from the d.c. supply whenever S_1 is closed. *Figure 2.5* shows a circuit that overcomes

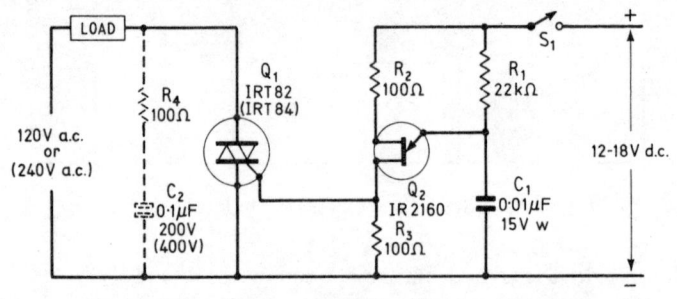

Figure 2.5. A.C. power switch with u.j.t. triggering.

this snag. Here, Q_2 is a free-running unijunction oscillator or pulse generator which feeds a series of high power trigger pulses into the triac gate whenever S_1 is closed. The oscillator operates at a frequency of several kHz, and generates roughly 50 trigger pulses during each half-cycle of the a.c. power line waveform. Consequently, the triac is fired by the first trigger pulse occurring in each half-cycle, and this pulse occurs within a few degrees of the start of the half-cycle. The triac is thus turned on almost permanently when S_1 is closed, and virtually full power is applied to the a.c. load.

The triac is turned on by pulsed (rather than steady) d.c. gate drive, and a considerable saving in gate drive current results. The u.j.t. circuit in fact draws a total current of only 4 mA or so from the d.c. supply. The d.c. supply can be derived from batteries, or from the a.c. power line via a suitable step-down transformer and rectifier and smoothing network.

A simple but very useful development of the *Figure 2.5* circuit is shown in *Figure 2.6*. This circuit is exactly the same as described above, except that the unijunction trigger pulses are fed into the triac gate via an isolating pulse transformer with a 1 : 1 turns ratio. Consequently, the u.j.t. circuit gives full on-off control of the triac, but is electrically fully isolated from the a.c. power line. The u.j.t. control circuitry can be grounded, for safety, if required.

15 A.C. POWER-SWITCHING PROJECTS 27

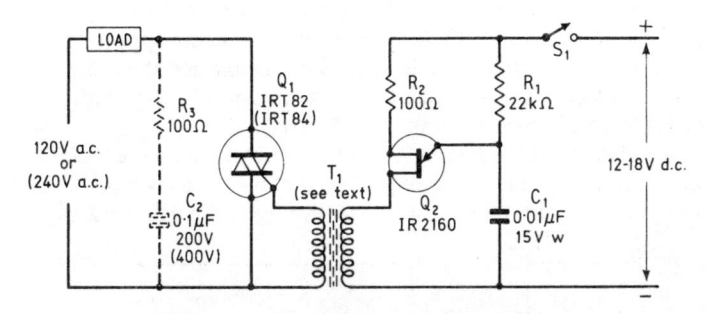

Figure 2.6. Isolated-input line switch.

Figure 2.7 shows how the isolated-input line switch can be modified to act as a water or steam-operated line switch that turns on in the presence of moisture. Q_3 is wired between the positive d.c. line and R_2 (the u.j.t.s main timing resistor), and Q_3 base is taken to a metal probe via limiting resistor R_1. When the probes are open zero base drive flows to Q_3, which thus acts as an open switch. Under this condition the u.j.t. fails to oscillate, and the triac is off. When a resistance less than a few megohms is placed across the probes, on the other hand, base drive flows to Q_3 via the negative supply line, and Q_3 acts like a closed

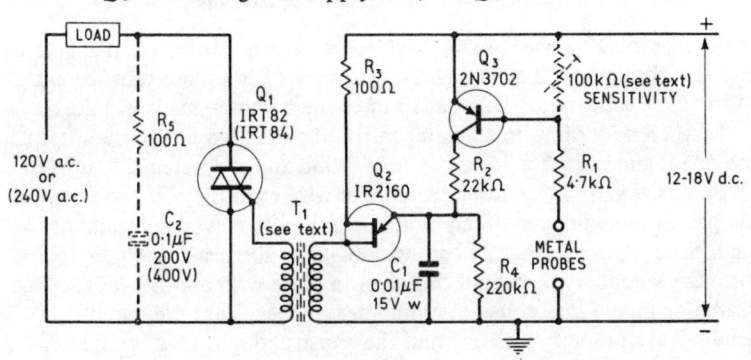

Figure 2.7. Water or steam-operated line switch (turns on in presence of moisture).

switch. Under this condition the u.j.t. oscillates in the normal way, and the triac turns on. Water and steam have distinct resistive qualities. Consequently, when water or steam come into contact with both metal probes simultaneously the triac turns on, and the circuit acts as a water or steam-operated line switch.

The action of the circuit can be reversed, so that the triac is normally on but turns off in the presence of water or steam, by using the connections shown in *Figure 2.8*. Q_3 in this circuit is wired across C_1. A shunt thus appears across the capacitor when moisture appears across the probes, and the u.j.t. then stops oscillating and the triac turns off. A resistance of less than two megohms must appear across the probes to ensure proper turn-off of the triac.

The sensitivities of the *Figure 2.7* and *2.8* circuits can be reduced to a pre-set level, if required, by wiring a 100 000 ohm pre-set resistor across the base-emitter junction of Q_3, as shown dotted in the diagrams. One

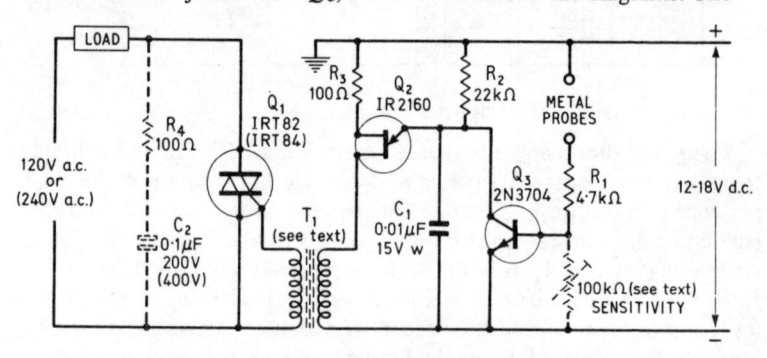

Figure 2.8. Water or steam-operated line switch (turns off in presence of moisture).

of the two metal probes used in each circuit can be grounded, if required, as is also shown in the diagrams. Transformer T_1 is a pulse transformer with a 1 : 1 turns ratio. This transformer can be home-made as follows.

Take a length of $3/8$ in diameter ferrite rod (as used in a.m. radio antennae) and cut off a piece 1 in long. Wind approximately 25 turns of 28 to 30 s.w.g. or a.w.g. insulated copper wire onto the piece to form the primary winding, and then bind in place with a layer of insulating tape. Now place a second 25-turn winding over the insulating tape to form the secondary, and bind this layer in place with another layer of insulating tape. Finally, use an ohmmeter to check that the two windings are properly isolated, and the construction is then complete.

Figure 2.9 shows another special type of power-switching circuit. In this case the circuit acts as a push-button operated self-latching line switch, and operates as follows. Suppose initially that the triac is off. Under this condition zero voltage appears across the load and, since the triac gate is at virtually the same potential as the MT_1 terminal, zero voltage appears on the triac gate or across the $C_1 - R_2$ network. The triac is thus off, C_1 is discharged, and zero power is developed in the load under this condition.

Suppose now that S_1 is momentarily closed. Under this condition a pulse of gate power is applied to the triac via R_1, and the triac turns on and power is applied to the load, causing C_1 to charge and discharge in response to the line voltage via R_2. The currents of C_1 are, however, close to 90° out of phase with the C_1 voltage. Consequently, when the line voltage falls to zero at the end of the half-cycle the C_1 discharge currents are at a peak value and flow directly into the triac gate and cause the triac to turn on again at the start of the following half-cycle.

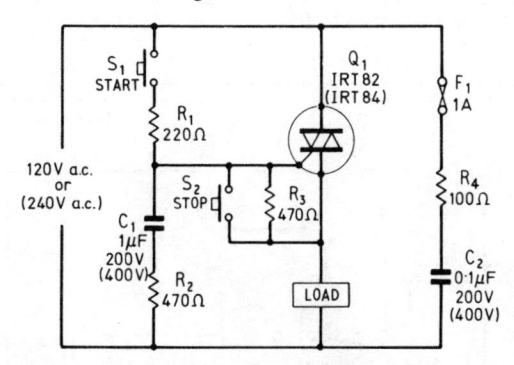

Figure 2.9. Push-button operated self-latching line switch.

R_2 and C_1 thus cause the triac to self-latch and stay on permanently once the operation has been initiated via S_1. The circuit can be turned off and reset by briefly closing S_2, causing C_1 to discharge through the load rather than the triac gate. A snubber network (R_4 and C_2) is wired across the power lines to prevent the triac being turned on by line transients.

Synchronous power-switching projects

All the projects that we have looked at so far generate a certain amount of r.f.i., either at the moment that power is first applied to the load or continuously so long as power is applied to the load. The magnitude of this r.f.i. is proportional to the magnitude of the instantaneous switching current of the triac, and may be quite objectionable when heavy-current resistive loads such as heaters are used, or when repetitive switching operations are carried out in lamp-flashing or similar applications using resistive loads.

These r.f.i. problems can be completely eliminated by using a synchronous 'zero-voltage' switching technique, in which gate drive is applied to the triac only in the brief periods when the instantaneous line voltage is close to the zero-voltage cross-over points of each half-cycle. *Figure 2.10* shows a practical synchronous on-off power switch of this type.

15 A.C. POWER-SWITCHING PROJECTS

Basically, the circuit contains a line-driven 'zero-crossing' detector section and a d.c.-powered triac-gate-drive section. These two sections are interconnected in such a way that a brief pulse of gate current is applied to the triac only when the line voltage is close to zero in each half-cycle. Q_2 and Q_3 form the zero-crossing detector, and are driven from the a.c. power line by current-limiting potential divider $R_2 - R_3$. The $Q_2 - Q_3$ connection is such that one or other of these transistors is

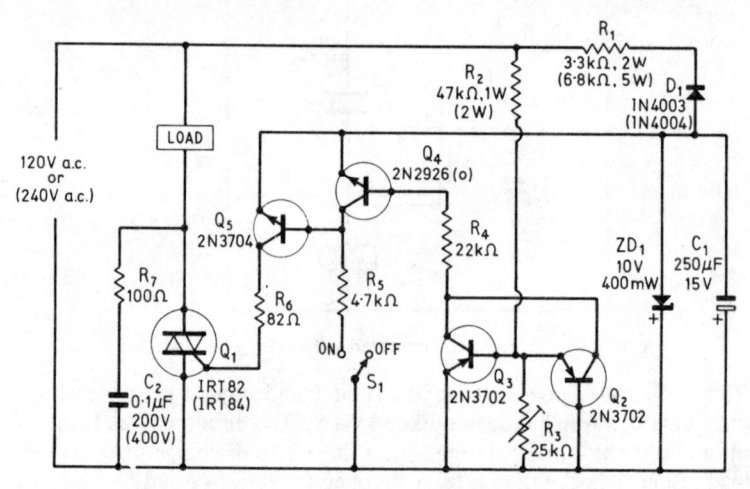

Figure 2.10. Zero-switching synchronous line switch (turns on with S_1 closed).

driven on whenever the line voltage exceeds a certain 'reference' value (adjustable down to 4·2 V via R_3), irrespective of the line voltage polarity, and so that both transistors turn off when the line voltage is below this reference value. The collectors of Q_2 and Q_3 are coupled to the base of Q_4 via R_4, and Q_4 uses R_5 as its collector load when S_1 is closed. R_5 also provides base drive to Q_5. Q_5 uses R_6 and the triac gate as its collector load. Q_4 and Q_5 are powered from a zener-regulated d.c. supply derived from the a.c. supply line via $R_1 - D_1 - ZD_1$ and C_1.

To understand the circuit action, assume that S_1 is closed and that the instantaneous a.c. line voltage is at some value in excess of the reference value of, say, 5 V. Under this condition either Q_2 or Q_3 is driven on and drives Q_4 to saturation via R_4. Q_4 has a saturation voltage that is lower than the base-emitter turn-on voltage of Q_5, so zero base drive flows to Q_5 under this condition, and Q_5 is cut off and zero gate drive is applied to the triac.

Suppose now that the instantaneous line voltage falls below the 5 V reference value, i.e., that the line voltage is very near the zero-voltage cross-over point near the end of one half-cycle and the start of the next. Under this condition Q_2 and Q_3 both turn off and remove the base drive of Q_4, which also turns off. Since Q_4 is cut off, heavy base drive flows to Q_5 via R_5, and Q_5 is driven to saturation. Heavy gate current thus flows into the triac gate via Q_5 and R_6 from the 10 V d.c. supply under this condition, and this current turns the triac on and causes it to self-latch for the duration of the half-cycle. Thus, gate trigger current is applied to the triac only in the brief periods when the line voltage is close to zero at the beginning and end of each half cycle, and negligible r.f.i. is generated under these switching conditions.

Note from the above description of circuit operation that gate drive can be applied to the triac only when base drive is available to Q_5 via R_5. The triac can thus be turned off or 'inhibited' by removing this base drive. The base drive can be removed in a number of ways. In *Figure 2.10* it is removed by wiring on-off switch S_1 in series with R_5, so the triac in this circuit turns on only when S_1 is closed. In *Figure 2.11*, on the other hand, it is removed by driving Q_4 permanently on via R_4, and in this case the triac turns on only when S_1 is open.

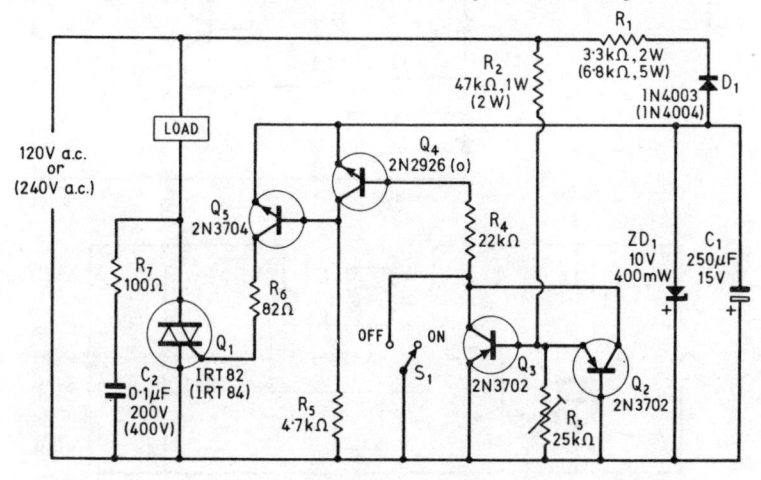

Figure 2.11. Zero-switching synchronous line switch (turns on with S_1 open).

The width of the gate pulse of the synchronous switching circuits can be varied via R_3. The width must be adjusted to ensure that the pulse does not end until the triac main-terminal currents have risen above the holding current in each half-cycle, otherwise the triac will fail to self-latch. To adjust R_3, connect the selected power load in place, set R_3 to maximum resistance, and apply power to the unit. Now reduce the R_3 value just past the point at which the triac turns on and

32 15 A.C. POWER-SWITCHING PROJECTS

applies power to the load. The voltage across C_1 should be checked when R_3 is adjusted, to ensure that it does not fall appreciably below the nominal 10 V value. The adjustment is then complete. If the triac is used with a multi-value load, the above adjustment must be made in the minimum load position only.

An integrated-circuit version of the synchronous zero-voltage switch is manufactured by General Electric, and is known as the PA424. *Figure 2.12a* shows the internal circuit of the device, and *Figure 2.12b*

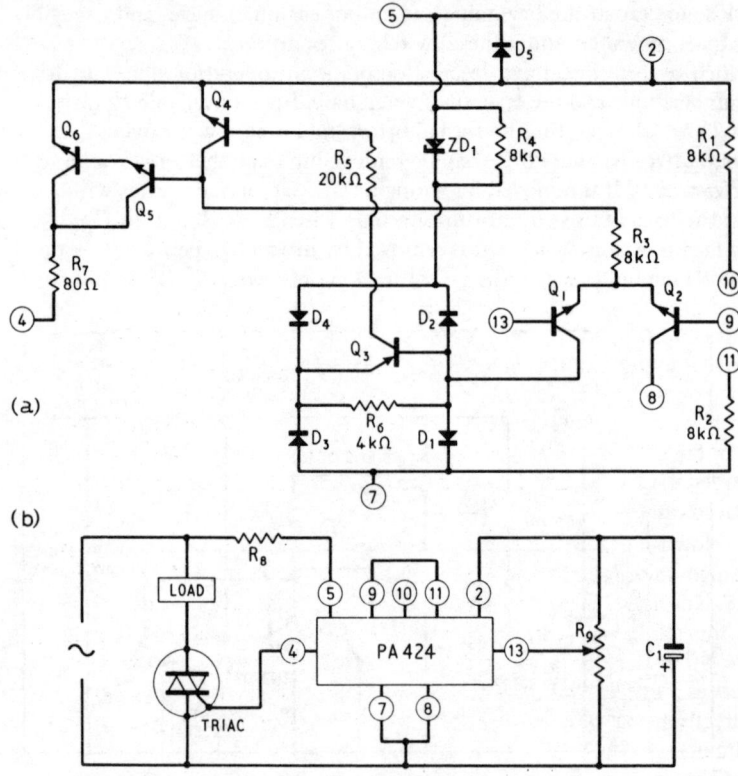

Figure 2.12a. Internal circuit of the PA424 integrated circuit.
Figure 2.12b. Basic connections of the PA424.

shows the basic method of connecting it to a triac. To understand the operation of the device, assume for the moment that $Q_1 - Q_2 - R_1 - R_2 - R_3$ and R_9 do not exist, and imagine that a 16 000 ohm resistor is wired between Q_4 collector and pin 7 of the i.c. Basic circuit operation is as follows.

The a.c. power line is normally connected between pins 5 and 7 of the i.c. via current-limiting resistor R_8, and an electrolytic smoothing

15 A.C. POWER-SWITCHING PROJECTS 33

capacitor is connected between pins 2 and 7. The peak value of the pin 5 alternating voltage is limited to about 10 V by zener diode ZD_1 and the series bridge-rectifier network formed by D_1 to D_4. The pin 5 voltage is rectified by D_5 and smoothed by C_1, so a stable d.c. supply is developed between pins 2 and 7. Q_3 is wired across the output of the a.c.-powered D_1 to D_4 bridge rectifier in such a way that it is driven on whenever the a.c. line voltage exceeds a few volts. Q_3 and the bridge rectifier thus act as a zero-crossing detector. Q_3 applies base drive to Q_4 via R_5 when Q_3 is on, and Q_4 uses the imaginary 16 000 ohm resistor mentioned above as its collector load. This resistor also provides base drive for Q_5 and Q_6, which are connected as a super-alpha pair and provide gate drive to the triac via R_7.

Thus, when the line voltage is appreciably above the zero-crossing point Q_3 is driven on and drives Q_4 to saturation. Q_4 removes all base drive from Q_5 under this condition, so Q_5 and Q_6 are cut off and zero gate drive is applied to the triac. When, on the other hand, the line voltage is very close to the zero-crossing point of each half-cycle Q_3 is cut off and applying zero base drive to Q_4, which is also cut off. Heavy base drive is applied to Q_5 via the imaginary 16 000 ohm resistor under this condition, so Q_5 and Q_6 are driven to saturation and heavy gate drive is applied to the triac via R_7. Thus, gate drive is applied to the triac only in the brief periods when the line voltage is close to the cross-over point in each half-cycle, and r.f.i.-free synchronous operation is obtained. Note at this point that the circuit can be inhibited or turned off by feeding a current into Q_3 base, so that Q_3 is permanently driven on.

Now that the basic operation of the circuit has been examined we can go on to look at the functioning of $Q_1 - Q_2 - R_1 - R_2 - R_3$ and R_9. Normally, pins 9, 10 and 11 of the i.c. are shorted together, pin 8 is shorted to pin 7, and pin 13 is taken to a tap on a potential divider (R_9) that is wired between pins 2 and 8. The $Q_1 - Q_2$ network thus acts as a differential amplifier, with Q_1 collector current feeding directly into Q_3 base. Q_1 can thus provide an 'inhibit' current to the circuit.

The action of the differential amplifier is such that Q_1 is fully cut off if R_9 is adjusted so that the pin 13 voltage is appreciably negative relative to the fixed voltage that is developed on pin 9 via R_1 and R_2. Under this condition Q_1 applies zero drive to Q_3, and the circuit gives synchronous operation of the triac. If, on the other hand, R_9 is adjusted so that pin 13 is positive to pin 9, Q_1 is driven on and passes an inhibit current into Q_3 base. Under this condition zero drive is applied to the triac, which is thus permanently off. A stable potential of roughly half the d.c. supply voltage appears on the emitters of Q_1 and Q_2 under all conditions. In the PA424, therefore, R_3 is wired between this point and the collector of Q_4, and serves both as a 'tail' for the differential amplifier and as a collector load resistor for Q_4 (in place of the imaginary 16 000 ohms resistor mentioned earlier).

Figures 2.13 and *2.14* show two practical synchronous power-switching circuits using the PA424. In both cases a potential divider is

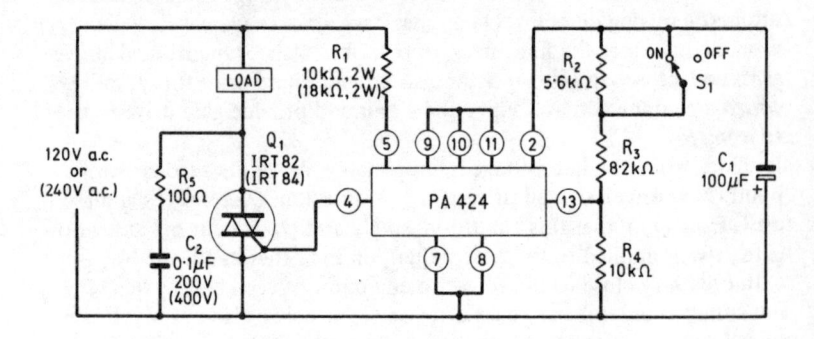

Figure 2.13. Zero-switching synchronous line switch using i.c. (turns on with S_1 closed).

wired between pins 2 and 7 so that a fixed voltage is developed on pin 13. This voltage can be shifted, so that the circuit is inhibited, via S_1. The *Figure 2.13* circuit is arranged so that the triac turns on when S_1 is closed, and the *Figure 2.14* circuit is arranged so that the triac turns on when S_1 is opened.

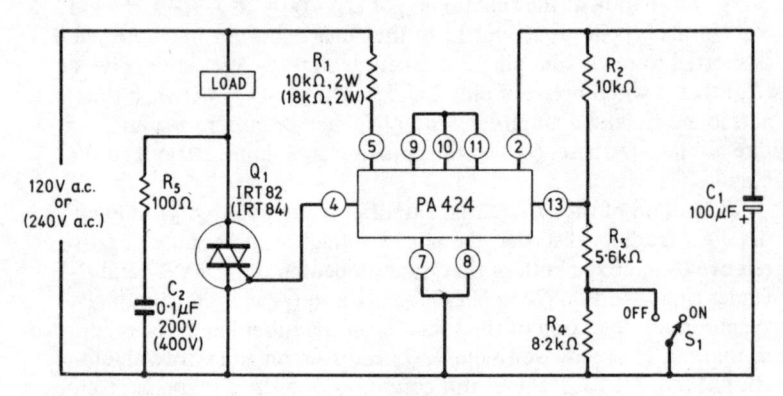

Figure 2.14. Zero-switching synchronous line switch using i.c. (turns on with S_1 open).

15 A.C. POWER-SWITCHING PROJECTS   35

Note in these circuits that the pulse width of the PA424 is not adjustable, and the triac will thus not self-latch if too low a value of triac load is used. The specified triacs will operate correctly with load powers down to a few hundred watts. If lower power loads are used, triacs with lower holding-current values must be employed.

Automatic overload-switching projects

Figures 2.15 and *2.16* show how the synchronous switching circuit of *Figure 2.10* can be modified to act as an overload switch that cuts out automatically if the load current rises above a pre-set value. In both circuits a low-value resistor, R_8, is wired in series with MT_1 of the triac. A voltage proportional to the load current is developed across this resistor, and this voltage is fed to an electronic cut-out on negative half-cycles via D_2.

In *Figure 2.15* the cut-out comprises $Q_6 - R_9 - R_{10} - R_{11}$ and C_3. The circuit action is such that C_3 charges to a voltage proportional to the peak load current on negative half-cycles. If C_3 charges above 5 V Q_6 is driven on via potential divider $R_9 - R_{10}$, and Q_6 drives Q_4 on via R_{11}, thus preventing the triac firing again at the end of the half-cycle.

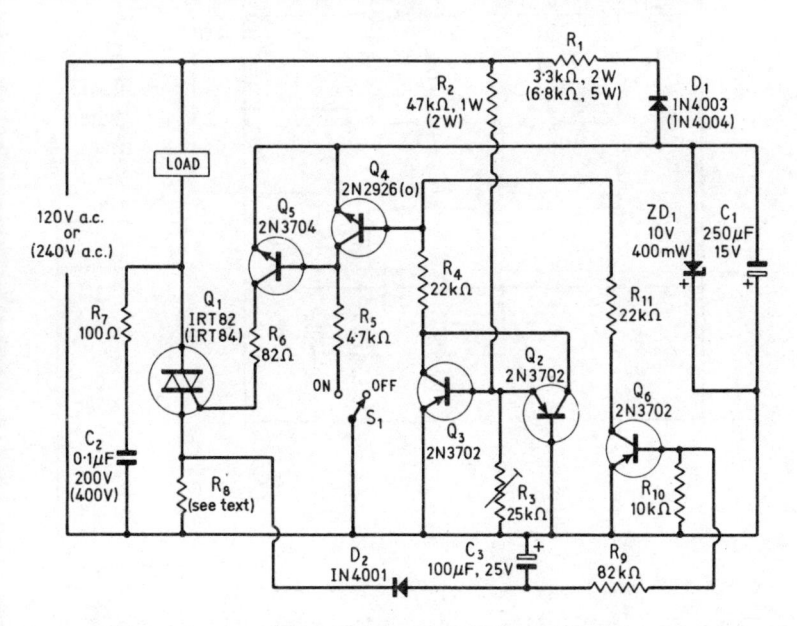

Figure 2.15. Zero-switching synchronous overload switch, non-latching.

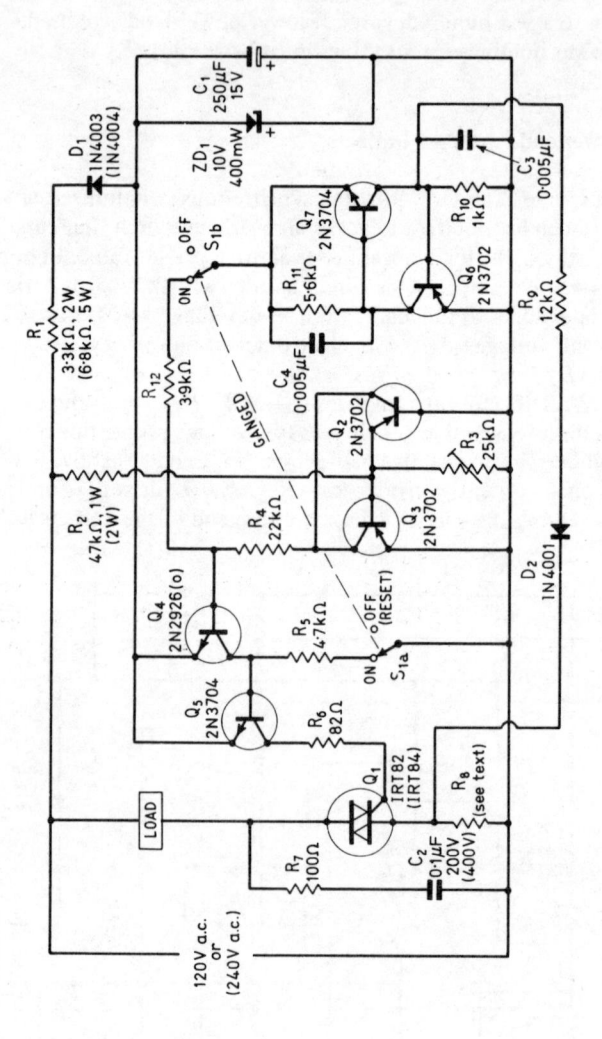

Figure 2.16. Zero-switching synchronous overload switch, self-latching.

The C_3 voltage then slowly decays, and when it falls below 5 V the triac fires again. If the load current is still in excess of the pre-set value the triac again cuts off after applying one or two half-cycles of power.

Thus, this overload switch does not latch into the 'trip' condition when the load current is exceeded, but gives a form of skip-cycling in which test cycles of power are applied to the load at periodic intervals. The period between test cycles is proportional to the magnitude of the overload current, and the r.m.s. power averaged over these periods is of a low order. Full power is applied to the load automatically once the overload is removed. The skip-cycling is not harmful to the triacs specified in this circuit, which have very high overload capabilities. Both triacs can safely handle non-repetitive one-cycle overload currents up to 80 or 100 A.

Figure 2.16 shows how the circuit can be modified to give a latching operation of the overload trip. Here, Q_6 and Q_7 are wired as a regenerative switch that turns on and self-latches when the peak R_8 voltage exceeds 5 V. This switch then removes the triac gate drive via R_{12} and Q_4. Once the circuit has tripped it can be reset again by sequentially turning S_1 off and on.

R_8 in these two circuits must be chosen so that it develops roughly 5 V at the required peak trip current; e.g., if the circuit is required to trip at 5 A peak, R_8 needs a value of one ohm, and if it is required to trip at 500 mA R_8 needs a value of 10 ohms. If these circuits are to be used with incandescent lamp loads, it should be noted that such lamps pass inrush or turn-on currents several times greater than their normal 'hot' running currents.

CHAPTER 3

20 ELECTRONIC ALARM PROJECTS

The high power gains, low leakage currents, and high current carrying capacities of s.c.r.s make them ideal for use in a variety of electronic alarm projects. A range of such projects are described in this chapter, and include both simple and advanced burglar alarms, light-beam alarms, smoke alarms, automatic fire alarms, over-temperature alarms, frost alarms, under-temperature alarms, and alarm circuits that are operated by contact with water or steam.

All of these projects are designed around the IR106Y1 s.c.r. manufactured by International Rectifier, and are reasonably versatile circuits. The alarm that each circuit uses can be any self-interrupting bell, buzzer, or siren that draws an operating circuit of less than 2 A. Each circuit must be operated from a battery supply roughly 1·5 V greater than the nominal operating voltage of the alarm device used.

When building these projects, check that the s.c.r. does not get too hot when the alarm is driven on. If it does, connect the s.c.r. heat tab to a heat sink with an area of a couple of square inches or greater.

Contact-operated alarm projects

The simplest type of alarm circuit that can be built around the IR106Y1 is the remote-operated alarm shown in *Figure 3.1*. The circuit is of the non-latching type, and operates when any of the input switches (S_1 to S_y) are closed. Dozens of these switches can be wired in parallel, and each will cause the alarm to operate. These switches pass *on* currents of only a few milliamps, so they can be placed hundreds of feet away from the s.c.r. and alarm, without risk of trouble from high cable resistance.

20 ELECTRONIC ALARM PROJECTS   39

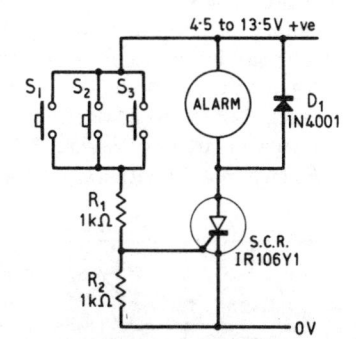

Figure 3.1. Remote-operated alarm.

The circuit can be converted to a simple self-latching burglar alarm or a multi-input fire alarm by wiring a latching resistor and reset button across the alarm, as shown in *Figure 3.2*.

These two circuits have plenty of uses in the home and in industry. They pass typical standby currents of only 0·1 μA when the alarm is in the *off* condition, so cause negligible drain on supply batteries. If the switches are of the microswitch or reed types, the alarms can be made to operate whenever a door or window is opened, or when an object travels beyond a pre-set limit. If the switches are of the pressure-pad type, the alarm can be made to operate whenever a person stands on a

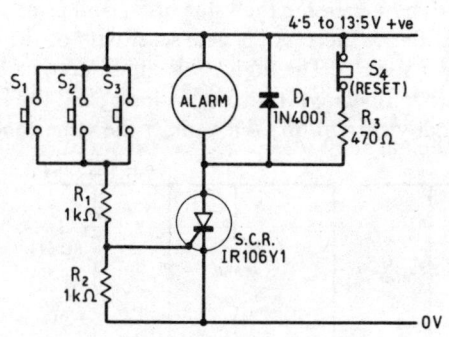

Figure 3.2. Simple make-to-operate burglar alarm, or multi-input fire alarm.

mat or a vehicle passes over a pressure-pad. If thermostat switches are used, the circuits act as automatic fire alarms.

The *Figure 3.2.* circuit is useful as a burglar alarm, but can be disabled by cutting the cable linking the switches to R_1 or to the positive supply line. A more reliable burglar alarm circuit is that shown in *Figure 3.3*, which operates and self-latches if any of the switch contacts are briefly operated, or if their connecting leads are cut.

40 20 ELECTRONIC ALARM PROJECTS

C_1 is a noise-suppressing capacitor, and ensures that the alarm will not be inadvertently operated by the action of switch contacts momentarily bouncing or sliding apart under vibration or shock conditions; the alarm will only operate if the contacts are held open for a period in excess of one millisecond or so.

The alarm circuit of *Figure 3.3* draws a standby current of 500 μA (via R_1) when used with a 6 V supply. The standby current can be

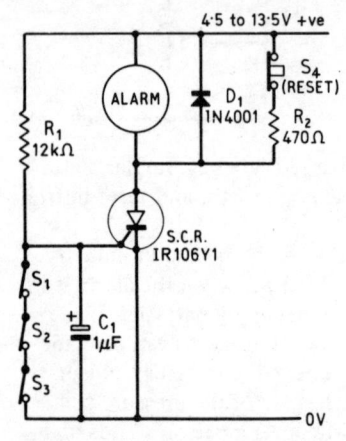

Figure 3.3. Break-to-operate burglar alarm. The circuit draws a standby current of 500 μA at 6 V.

reduced, if required, by increasing the value of R_1 and using a one or two-transistor amplifier to increase the gate sensitivity of the s.c.r., as shown in *Figures 3.4* and *3.5*. The *Figure 3.4* circuit draws a standby current of 60 μA when operated from a 6 V supply, and the *Figure 3.5* circuit draws a standby current of 1·4 μA under the same condition.

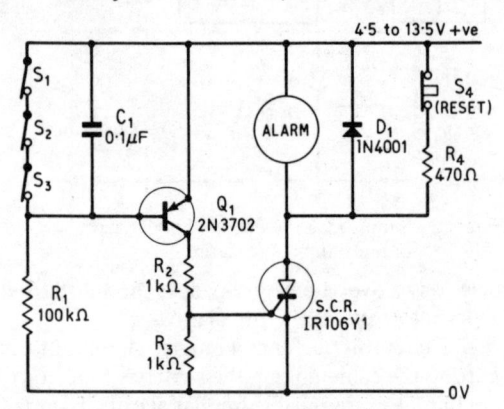

Figure 3.4. Break-to-operate burglar alarm, drawing a standby current of 60 μA at 6 V.

The break-to-operate burglar alarms of *Figures 3.3* to *3.5* are far more useful than the simple-make-to-operate type of *Figure 3.2*, but are still not fully tamper proof. They can be disabled by connecting a jumper lead across the normally closed switch contacts. This snag can be overcome by combining both break-to-operate and make-to-operate

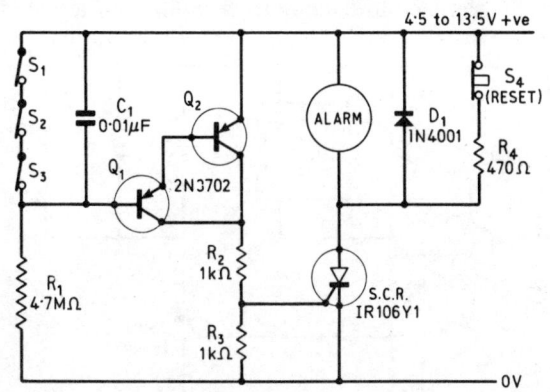

Figure 3.5. Break-to-operate burglar alarm, drawing a standby current of $1.4\,\mu A$ at 6 V.

switches in a single alarm, as shown in the circuits of *Figures 3.6* and *3.7*. A burglar is unlikely to know which alarm leads are of the break or make types, and if he cuts or shorts the wrong ones the alarm will instantly sound and self-latch.

The *Figure 3.6* tamper-proof alarm is developed by combining the circuits of *Figures 3.2* and *3.3*, and draws a standby current of 500 μA

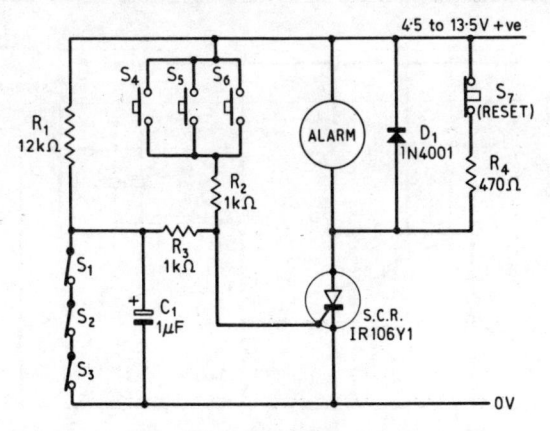

Figure 3.6. Simple 'tamper-proof' burglar alarm, drawing a standby current of 500 μA at 6 V.

42 20 ELECTRONIC ALARM PROJECTS

from a 6 V supply. The *Figure 3.7* circuit is developed by combining the circuits of *Figures 3.2* and *3.5*, and draws a standby current of 1·4 µA from a 6 V supply.

All the burglar alarm circuits looked at so far turn on and self-latch as soon as any of the input switches are operated. Consequently, if the owner turns the alarm on to standby from within a protected building

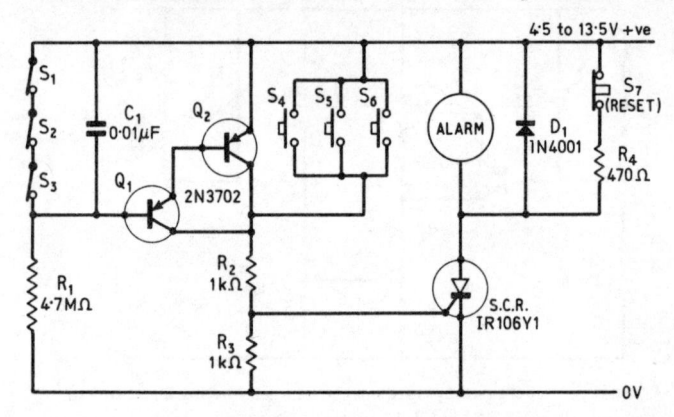

Figure 3.7. High-performance 'tamper-proof' burglar alarm, drawing a standby current of 1·4 µA at 6 V.

it is impossible for him to leave that building without causing the alarm to operate and self-latch. This snag can be overcome by fitting the alarm with a delayed self-latching facility, which ensures that the alarm does not go into the self-latching mode until some fixed time after the system is initially put into the standby condition. *Figure 3.8* shows the practical

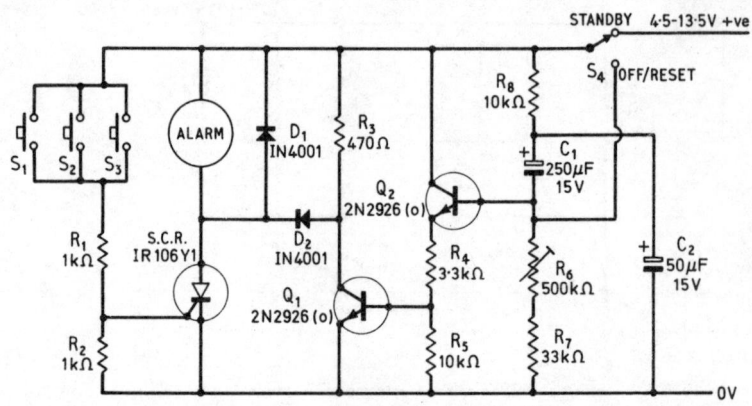

Figure 3.8. Simple burglar alarm with delayed self-latching facility giving the owner time to leave a secured building without sounding the alarm.

circuit of a delayed self-latching facility connected up to a simple alarm system of the type shown in *Figure 3.2*. The facility can, however, be added to any one of the other alarm circuits already described, since the $D_2 - R_3 - Q_1 - Q_2$ – etc. circuit is used as a time switch that is used to replace the latching resistor and reset button of the earlier circuits.

The operation of the *Figure 3.8* circuit is fairly simple. C_1 and $R_6 - R_7$ form a time-controlled potential divider that is connected to the base of emitter follower Q_2. The emitter current of Q_2 feeds into the base of common emitter amplifier Q_1, which uses R_3 as its collector load. R_3 is connected to the s.c.r. anode via D_2. R_8 and C_2 act as a simple filter network that provides smooth d.c. to the C_1 and R_6-R_7 timing network.

When power is first applied to the circuit by moving S_4 to the standby position C_1 is fully discharged, and Q_2 base is thus effectively shorted to the positive supply line. Heavy base drive is thus applied to Q_1 via Q_2, and Q_1 is driven to saturation. The $R_3 - D_2$ junction is thus effectively shorted to the zero volts line under this condition, so the s.c.r. will not self-latch if any of the input switches are closed at this time.

As soon as power is applied to the circuit via S_4, C_1 starts to charge exponentially via R_6 and R_7, and the voltages on the base and emitter of Q_2 start to decay slowly towards zero. Eventually, after a period between a few seconds and a couple of minutes (depending on the setting of R_6), the emitter current of Q_2 falls to such a low value that Q_1 comes out of saturation and eventually turns off, at which point it draws zero current from R_3. If any of the alarm switches are operated at or after this time, therefore, the s.c.r. and alarm will turn on and self-latch via R_3 and D_2.

Thus, the delayed latching facility enables the owner to turn the alarm system on to standby from within a protected building, and then to leave that building via a protected doorway. As the owner passes through the protected door the alarm sounds, indicating that the system is fully functional, but the alarm ceases to sound as the owner closes and locks the door behind him. A short time later the alarm system automatically connects the self-latching facility, and the system then operates in the normal way if any of the alarm switches are subsequently operated.

An alternative device that can be used to enable the owner to leave a protected building is the relay time switch. This device is connected across one of a building's exit doors, and is arranged so that it disables the door alarm switch for a brief period as the owner leaves the building. The device lets the owner leave the building without sounding the alarm.

Figure 3.9 shows the practical circuit of a relay time switch. The circuit is almost identical to that of the delayed latching facility of *Figure 3.8*, except that R_3 of *Figure 3.8* is replaced by relay RLA in *Figure 3.9*. Circuit operation is initiated by briefly closing S_1, thus

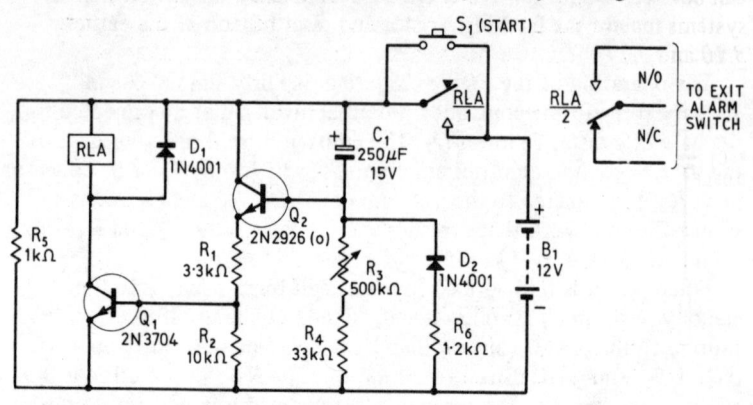

Figure 3.9. Relay time-switch. This circuit can be used to enable the owner to leave a protected building without sounding the alarm.

RLA = any 12 V relay with a coil resistance of 180 or greater, and with two or more sets of change-over contacts.

driving RLA and Q_1 on via Q_2 and C_1; as the relay goes on contacts $RLA/1$ change over and bypass S_1, thus maintaining power to the circuit when S_1 is released, and contacts $RLA/2$ change over and disable the alarm switch in the exit door. C_1 starts to charge via R_3 and R_4 as soon as S_1 is closed; eventually, after a delay variable between 10 seconds and 2 minutes via R_3, the C_1 charge rises to such a level that RLA and Q_1 turn off; as RLA turns off contacts $RLA/1$ change over and remove all power from the circuit, and contacts $RLA/2$ change over and enable the exit doors alarm switch to operate in the normal way. Once power is removed from the circuit by the turn-off of RLA C_1 discharges rapidly via $R_5 - R_6$ and D_2. The operating sequence of the circuit is then complete. Relay switches $RLA/2$ can be used to disable switches of both the break-to-operate and make-to-operate types.

Water and steam-operated alarms

The impurities in normal water cause the liquid to act as a conductive medium that exhibits the characteristics of a resistor. Many other liquids, and vapours such as steam, exhibit similar resistive qualities.

Consequently, the resistance across a pair of isolated metal probes falls from near infinity to some moderately low value whenever the probes are placed together in these conductive media. These resistance changes can be used to activate a variety of electronic alarm systems. Such systems may be used as water or steam-operated alarms, and *Figures 3.10* and *3.11* show two practical circuits of these types.

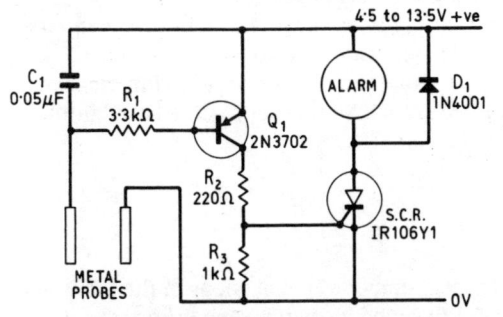

Figure 3.10. Water-operated alarm.

Both of these circuits are operated in the non-latching mode, and use a transistor amplifier to effectively increase the s.c.r. gate sensitivity to such a degree that the s.c.r. is triggered by the small electrical currents that pass through the conductive medium when it comes into contact with both probes simultaneously.

In the *Figure 3.10* circuit a one-transistor amplifier is used, and the circuit sensitivity is such that the alarm is driven on whenever a resistance of less than 220 k ohms appears across the probes. This sensitivity is sufficient to cause the circuit to act as a water-operated alarm.

The *Figure 3.11* circuit used a two-transistor amplifier, and the circuit sensitivity is such that the alarm is driven on whenever a resistance of less than 10 megohms appears across the probes. This

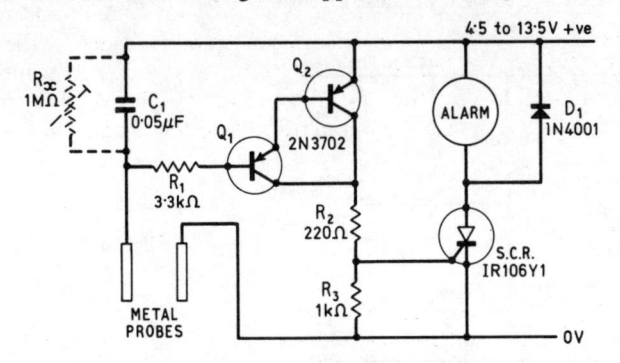

Figure 3.11. Steam-operated alarm.

sensitivity is sufficient to cause the circuit to act as either a water or steam-operated alarm. The circuit sensitivity can be reduced to a preset level, if required, by wiring a one-megohm pot across C_1, as shown dotted in the diagram. C_1 in these two circuits is used to suppress any a.c. pick-up from long connecting leads, which might otherwise cause the circuits to operate erratically, and R_1 is used to protect the circuits in the event of a short being placed directly across the probes.

These two circuits have a variety of applications in the home and in industry. They can be used to sound an alarm when it rains, when flooding occurs in cellars, when water rises to a preset level in tanks or baths, or when steam is ejected from a kettle spout as the liquid in the kettle starts to boil.

Light-operated alarm projects

Light-operated alarms have a number of applications in the home and in industry. They can be used to sound an alarm when light enters a normally dark area, such as the inside of a safe or strong-room, or they can be used to sound an alarm if an intruder or object breaks a projected light beam. They can also be used as smoke-sensitive alarms.

A variety of light-operated alarms are described in this section, and all use an l.d.r. as a light-sensing element. This light-dependent resistor is a cadmium-sulphide photocell, and acts as a high resistance under dark conditions and as a low resistance when brightly illuminated. All these circuits are versatile types, and will work with almost any l.d.r.s with face diameters in the range $1/8$ in to $1/2$ in; no precise l.d.r. types are thus specified in these circuits; notes on l.d.r. selection are, however, given where applicable.

Figures 3.12 and *3.13* show the circuits of two simple light-operated alarms. The l.d.r. in each of these circuits is meant to be mounted in a

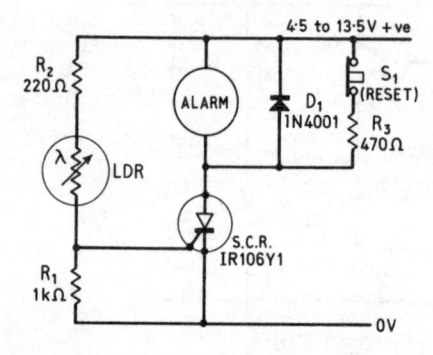

Figure 3.12.

Simple light-activated alarm.

normally dark area such as a safe or strong-room, and the designs are such that the alarm sounds when a light is shone into the protected area. The l.d.r. and R_1 form a potential divider that supplies gate drive to the s.c.r. Gate drive is supplied directly in the *Figure 3.12* circuit, and via an emitter-follower in the *Figure 3.13* circuit.

Circuit operation is very simple. Under dark conditions the l.d.r. presents a high resistance, so zero drive is applied to the s.c.r. gate. When the l.d.r. is illuminated its resistance falls to a low value, and gate drive is then applied to the s.c.r., which turns on and activates the

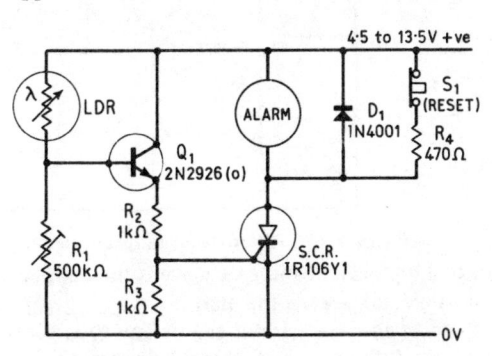

Figure 3.13. Sensitive light-activated alarm.

alarm. The sensitivity of the *Figure 3.12* circuit is such that the alarm turns on when the l.d.r. resistance falls to less than about 10 000 ohms. The sensitivity of the *Figure 3.13* circuit is much higher, and this design turns on when the l.d.r. resistance falls to less than 200 000 ohms. the sensitivity of the *Figure 3.13* circuit can be reduced via R_1.

The operation of the *Figure 3.12* and *3.13* circuits can be reversed, so that the alarm sounds when the l.d.r. illumination is decreased, by simply transposing the R_1 and l.d.r. positions as shown in *Figures 3.14*

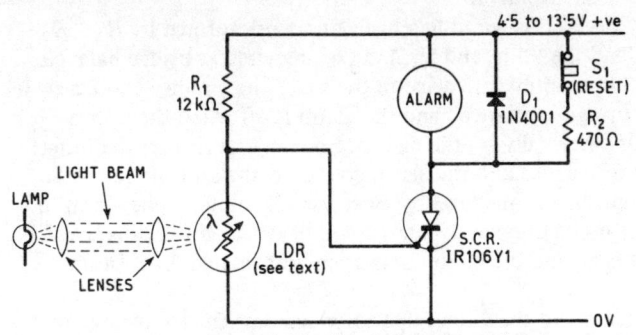

Figure 3.14. Simple light-beam alarm.

and *3.15*. These two circuits can be used as simple interrupted-light-beam alarms. Normally, the l.d.r. is brightly illuminated via a projected light-beam and lens system. The l.d.r. presents a low resistance under this condition, and zero drive is applied to the gate of the s.c.r., which is thus off. When a person or object enters the light beam the l.d.r.

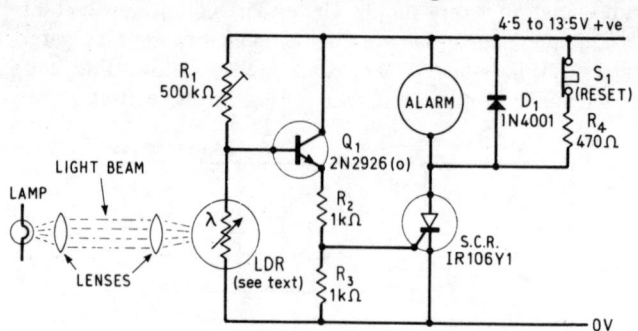

Figure 3.15. Alternative light-beam alarm.

resistance rises to a high value, and gate drive is then applied to the s.c.r., which turns on and sounds the alarm.

The l.d.r. in the *Figure 3.14* circuit can be any type that offers a resistance less than 1 000 ohms under the illuminated condition, and more than 3 000 ohms under the 'interrupted' condition. The *Figure 3.15* circuit can be used with any l.d.r. that offers a resistance less than about 50 000 ohms under the illuminated condition.

The circuits of *Figures 3.14* and *3.15* act as useful intrusion alarms in many applications, but are not suitable for use as high-security burglar alarms. Both circuits can be disabled by shining a bright light, with an intensity greater than that of the normal light-beam, on to the l.d.r. face. This vulnerability of the basic light-beam alarm can be overcome in a number of ways. *Figure 3.16* shows one circuit that overcomes the problem.

Here, the l.d.r. is wired in a bridge network formed by $R_1 - R_2 - R_3 - R_4$ and the l.d.r., and Q_1 and Q_2 are wired as bridge balance detectors that apply gate drive to the s.c.r. The circuit action is such that the bridge is balanced and the alarm is off when the l.d.r. is illuminated normally by the light beam, and is such that the bridge goes out of balance and the alarm goes on if the l.d.r. illumination varies from the normal level by even a small amount. The alarm thus goes on if the light-beam is interrupted or if a bright light is shone on the l.d.r. face, and the circuit acts as an 'unbeatable' light-beam alarm.

To understand the circuit operation, assume the following points. The circuit is powered from a 10 V supply. R_1 is adjusted so its resistance is equal to that of the l.d.r. in the normal 'balanced' condition in which the l.d.r. is illuminated by the light-beam, so 5 V is developed on Q_1 base and Q_2 emitter under this condition. R_3 is adjusted so that 5·6 V is developed on Q_1 emitter, and 4·4 V is developed on Q_2 base, i.e., a forward bias of 600 mV is developed

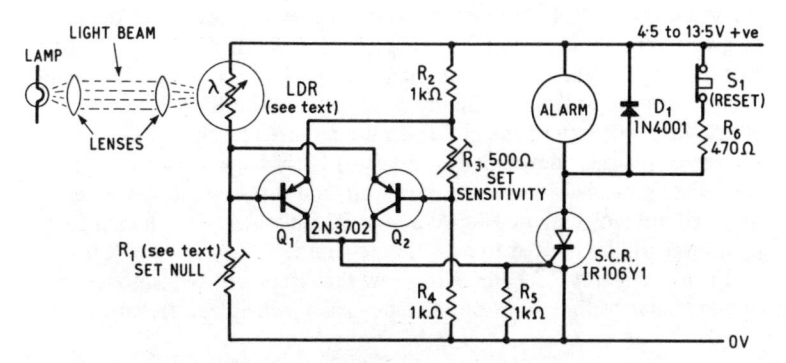

Figure 3.16. 'Unbeatable' light beam alarm.

between the base-emitter junctions of Q_1 and Q_2 under the normal balanced condition. Q_1 and Q_2 each turn on when the forward base-emitter bias rises to 650 mV.

Thus, under the normal balanced condition a forward bias is applied to both Q_1 and Q_2, but is not sufficient to cause either transistor to conduct, so the s.c.r. and alarm are off. Suppose, however, that the light-beam is interrupted, so that the l.d.r. resistance increases. Under this condition the voltage at the $LDR-R_1$ junction falls appreciably below 5 V, so the Q_1 forward base-emitter bias rises above 650 mV, and Q_1 conducts and drives on the s.c.r. and the alarm.

Alternatively, suppose that a bright light, with an intensity greater than that of the basic light-beam, is shone on the l.d.r. face, so that the l.d.r. resistance decreases. In this case the voltage at the $LDR-R_1$ junction rises appreciably above 5 V, so the Q_2 forward base-emitter bias rises above 650 mV, and Q_2 conducts and again drives on the s.c.r. and the alarm.

The circuit thus operates the alarm if the l.d.r. illumination varies from the normal light-beam level by an amount sufficient to cause a voltage change greater than 50 mV at the $LDR-R_1$ junction. The sensitivity of the circuit can be varied via R_3, and can be set to such high levels that the alarm can be activated by changes in light level too small to be detected by the human eye.

The l.d.r. used in the *Figure 3.16* circuit can be any type giving a resistance in the range 200 ohms to 2 000 ohms when illuminated by the light-beam. R_1 should have a maximum value roughly double that of the l.d.r. under the above condition. To set up the circuit, proceed as follows.

First, adjust R_1 so that roughly half-supply voltage is developed at the $LDR-R_1$ junction when the l.d.r. is illuminated, and then adjust R_3 so that roughly 400 mV is developed across R_5. Now readjust R_1 to give a minimum reading across R_5; readjust R_3, if necessary, so that this reading does not fall to less than 200 mV. When the R_1 adjustment is complete the bridge is correctly balanced. R_3 can then be adjusted to set the sensitivity of the circuit to the required level. If R_3 is set so that zero voltage is developed across R_5 a fairly large change in light level will be needed to operate the alarm, and if it is set so that a few hundred millivolts are developed across R_5 only a very small change in light level will be needed to operate the alarm.

Finally, *Figures 3.17* and *3.18* show the circuits of two smoke-operated alarms that work on the light-beam principle. Here, the l.d.r.

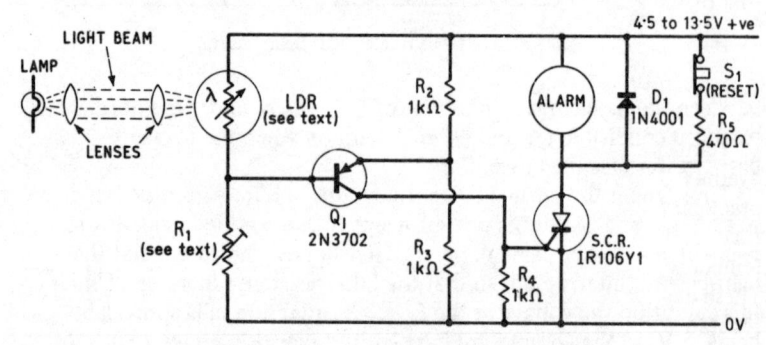

Figure 3.17. Simple smoke-operated alarm.

is illuminated by a light-beam and is wired in a bridge network formed by $R_1 - R_2 - R_3$ and the l.d.r., and the bridge output is detected and used to provide gate drive to the s.c.r. A one-transistor detector is used in the *Figure 3.17* circuit, and a two-transistor differential detector is used in the *Figure 3.18* circuit.

Each circuit is adjusted so that the bridge is close to balance when the l.d.r. is illuminated normally, and Q_1 is conducting slightly but not sufficiently to drive the s.c.r. on. When smoke enters the light-beam the l.d.r. illumination decreases and the l.d.r. resistance increases, and the bridge then goes out of balance in such a way that Q_1 is driven further on, thus causing the s.c.r. and the alarm to turn on. The circuits thus function as smoke-operated alarms.

The l.d.r. used in the smoke-alarm circuits can be any type with a resistance in the range 200 ohms to 10 000 ohms when illuminated by

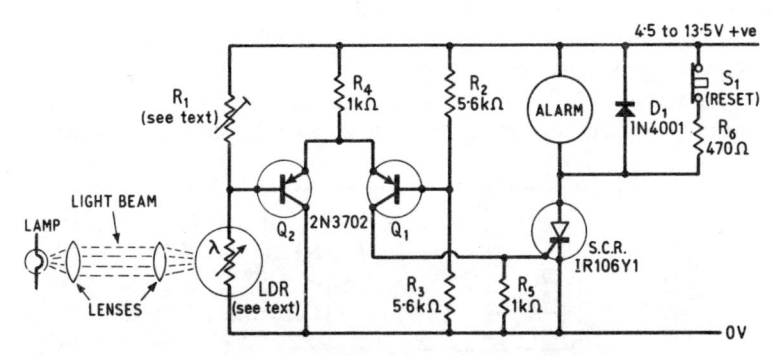

Figure 3.18. Sensitive smoke-operated alarm.

the light-beam. R_1 should have a maximum resistance roughly double that of the l.d.r. under the above condition.

Temperature-operated alarms

Temperature-operated alarms can be used as automatic fire or overheat alarms, or as frost-warning or underheat alarms. Three practical temperature-operated alarm circuits are described in this final section of this chapter. These circuits use inexpensive negative temperature coefficient thermistors as temperature-sensing elements. These devices act as temperature-sensitive resistors that present a high resistance at low temperatures, and a low resistance at high temperatures.

The three circuits described here are designed to work with thermistors that present a resistance of roughly 5 000 ohms at the desired operating temperature. The designs are quite versatile, however, and will in fact operate with any n.t.c. thermistors that present a resistance in the range 1 000 ohms to 20 000 ohms at the required temperature.

Figure 3.19 shows the practical circuit of a simple fire or over-temperature alarm. $R_1 - R_2 - R_3$ and thermistor TH_1 are wired in the form of a bridge, and Q_1 is used as a bridge balance detector and s.c.r. driver. R_1 is adjusted so that the bridge is balanced at a temperature just below the required operating level. Q_1 base and emitter are at equal potentials under this condition, so Q_1 and the alarm are off. As the TH_1 temperature rises the TH_1 resistance falls, so Q_1 base becomes negative to the emitter. When TH_1 reaches the required operating temperature Q_1 base becomes so negative that the transistor turns on and applies gate drive to the s.c.r., which then turns on and activates the alarm. The circuit thus operates as an alarm that turns on when the temperature exceeds a preset level.

The action of the circuit can be reversed, so that the alarm turns on when the temperature falls below a pre-set level, by simply transposing

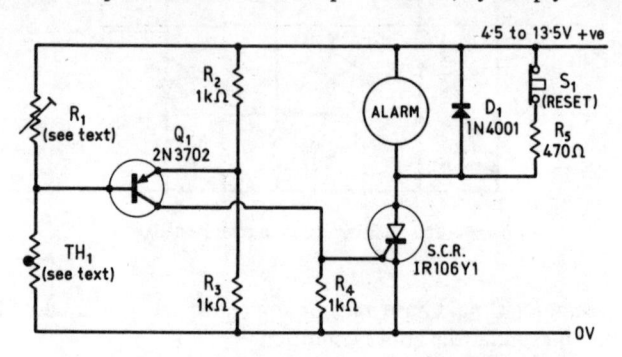

Figure 3.19. Simple fire or over-temperature alarm.

the R_1 and TH_1 positions, as shown in *Figure 3.20*. This circuit can be used as a frost or under-temperature alarm.

The circuits of *Figures 3.19* and *3.20* perform very well as temperature alarms, but their precise operating points are subject to slight variation with changes in Q_1 temperature, due to the temperature dependence of the characteristics of the Q_1 base-emitter junction. The circuits are thus not suitable for use in precision applications, unless Q_1 and TH_1 are

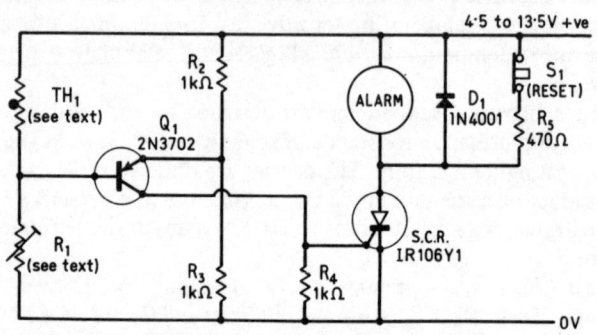

Figure 3.20. Simple frost or under-temperature alarm.

operated at the same temperatures. This snag can be overcome by using a two-transistor differential detector in place of Q_1, as shown in *Figure 3.21*.

This circuit is wired as a precision over-temperature alarm. It can be made to function as a precision under-temperature alarm by simply transposing the R_1 and TH_1 positions. Note that the circuit is shown without a latching resistor. In most applications of a sensitive circuit of this type the alarm is required to sound only so long as the preset temperature is exceeded.

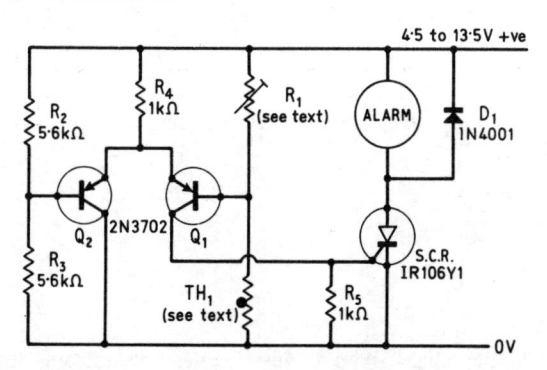

Figure 3.21. Precision over-temperature alarm.

TH_1 in these three circuits should, as already mentioned, be selected so that it has a resistance in the range 1 000 to 20 000 ohms at the required operating temperature. R_1 should be chosen so that it has a maximum resistance roughly double that of TH_1 under the above condition.

CHAPTER 4

15 TIME-DELAY PROJECTS

S.C.R. and triac switches can be readily coupled to electronic time-delay circuitry and used in a multitude of applications in the home, in industry, and in automobiles. These time-delay switches can be made to function in either the delayed turn-on or the automatic-turn-off modes, and can be used to control either a.c. or d.c. loads.

D.C. delayed turn-on projects

One of the simplest types of time-delay circuit is the delayed turn-on d.c. switch shown in *Figure 4.1*. Here, the s.c.r. is normally off, but turns on automatically some fixed time after switch S_1 is closed.

Circuit operation is very simple. When S_1 is first closed the s.c.r. is off, so negligible power is developed in the load and almost the full supply potential appears across the s.c.r. and the Q_1 u.j.t. time-delay

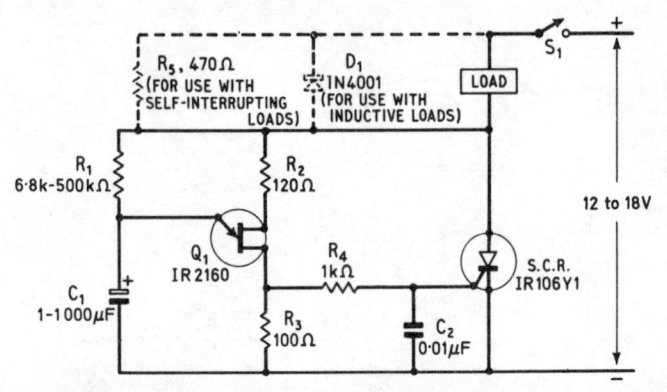

Figure 4.1. Delayed turn-on d.c. switch, giving delays up to 15 minutes.

network, and C_1 immediately starts to charge exponentially via R_1. After a preset delay the C_1 potential rises to the firing voltage of the u.j.t., which then fires and applies a trigger pulse to the gate of the s.c.r., which turns on. As the s.c.r. goes on it applies power to the load, and the s.c.r. anode falls to near-zero volts, thus effectively removing power from the u.j.t. timer network. The operating cycle is then complete.

R_1 in this circuit can be given any value in the range 6·8k to 500k ohms, and C_1 can have any value up to 1 000 μF. C_1 must, however, be a low-leakage specimen if large R_1 values are used. The time-delay of the circuit is approximately equal to the $R_1 - C_1$ product, where R_1 is given in kilohms, C_1 is given in μF, and t is given in milliseconds. Thus, values of 500k ohms and 100 μF give a delay of roughly 50 seconds. In practice, the delay using these component values usually works out at about 80 seconds, due to the fact that the tolerances of electrolytic timing capacitors are such that the actual C_1 value is normally about 60 per cent greater than its marked value.

The *Figure 4.1* circuit can be used to give maximum time delays up to about 15 minutes. This circuit, and all other d.c. time-delay circuits using the IR106Y1 as a switching element, can be used with any type of load up to a maximum current of 2 A. If an inductive load is used, diode D_1 (shown dotted) must be wired across the load as indicated. If a self-interrupting load is used, R_5 must also be wired across the load. Capacitor C_2 is wired between the gate and cathode of the s.c.r. to protect the s.c.r. against unwanted turn-on via line transients, which may reach the s.c.r. gate via the u.j.t. network.

Figure 4.2 shows a transistor version of the delayed turn-on d.c. switch, giving maximum delays up to several minutes. Circuit operation is fairly simple. R_1 and C_1 are wired across the supply as a time-controlled potential divider that applies a variable potential to Q_1 base via D_1, and $R_2 - R_3$ are wired across the supply as a potential divider

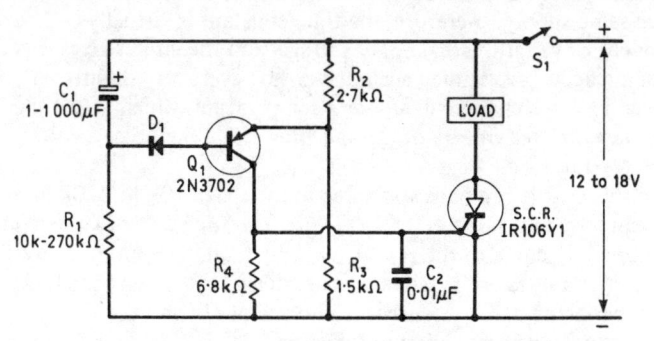

Figure 4.2. Delayed turn-on d.c. switch, giving delays up to several minutes. D_1 = general purpose silicon diode.

that applies a fixed reference voltage equal to roughly 3/10 of the supply line voltage to Q_1 emitter. Q_1 collector is connected to the gate of the s.c.r. The circuit action is such that the base-emitter junction of Q_1 is reverse biased when the $R_1 - C_1$ potential is above the $R_2 - R_3$ reference voltage value, and under this condition Q_1 and the s.c.r. are off. When, on the other hand, the $R_1 - C_1$ potential is below the reference value, the base-emitter junction of Q_1 becomes forward biased, and under this condition Q_1 and the s.c.r. turn on. Diode D_1 is used to protect the base-emitter junction of Q_1 against breakdown when it is heavily reverse biased.

Thus, when power is first applied to the circuit C_1 is fully discharged and acts like a short circuit, so Q_1 base is effectively shorted to the positive supply line, and Q_1 and the s.c.r. are off. C_1 starts to charge exponentially via R_1 as soon as S_1 is closed, and the C_1 voltage slowly increases. Eventually, after a preset delay, the C_1 voltage rises to such a degree that the Q_1 base voltage falls below the fixed $R_2 - R_3$ reference voltage, and the Q_1 base-emitter junction becomes forward biased and Q_1 turns on and applies gate drive to the s.c.r., which turns on and self-latches. The circuit action is then complete. Once the s.c.r. has turned on it can be turned off again by opening S_1. C_1 then discharges rapidly via $R_2 - D_1$ and the base-emitter junction of Q_1, and the circuit is then ready to operate again the next time S_1 is closed.

There are two points of particular note about this circuit. The first point is that Q_1 takes no current from the $R_1 - C_1$ network until the Q_1 turn-on voltage is reached, so the time constant of the unit is dictated purely by the R_1 and C_1 values, and is not noticeably influenced by the transistor characteristics.

The second point is that the s.c.r. turns on as soon as the $R_1 - C_1$ voltage falls below the $R_2 - R_3$ reference potential, which is a fixed fraction of the supply line voltage. Since both potential dividers are fed from the same supply, therefore, the time constant is virtually independent of variations in supply voltage, and the circuit gives very good timing accuracy. Timing accuracy is, however, very slightly influenced by large changes in ambient temperature, which affect the forward base-emitter voltage of Q_1 and thus affect the precise voltage at which Q_1 turns on.

R_1 in this circuit can have any value in the range 10k to 270k ohms, and C_1 can have any value up to 1 000 µF. The time delay of the circuit is approximately equal to the $R_1 - C_1$ product, and maximum delays up to several minutes are possible. Longer delays can be obtained by using a super-alpha pair of transistors in place of Q_1, and then increasing the R_1 value, as shown in *Figure 4.3*.

R_1 in this circuit can have a value up to a maximum of about 4·7 megohms, and delays up to about one hour are possible. C_1 must,

15 TIME-DELAY PROJECTS 57

however, be a reasonably low-leakage capacitor if very large R_1 values are used, otherwise the leakage-resistance potential divider action with R_1 may prevent the capacitor voltage from rising to the turn-on voltage of Q_1.

The *Figure 4.2* and *4.3* circuits act as excellent timers, but their accuracy is slightly susceptible to changes in ambient temperature. This

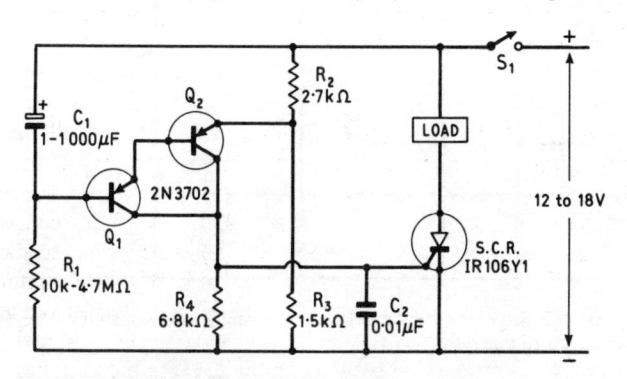

Figure 4.3. Long-period delayed turn-on d.c. switch, giving delays up to one hour.

snag can be overcome by using differential voltage detectors in place of the simple detectors already described. *Figures 4.4.* and *4.5* show two precision time-delay circuits using this modification.

The *Figure 4.4* circuit uses a two-transistor differential detector, and gives precision delays up to several minutes. D_1 and D_2 are used to protect the base-emitter junctions of the transistors against breakdown when they are heavily reverse biased.

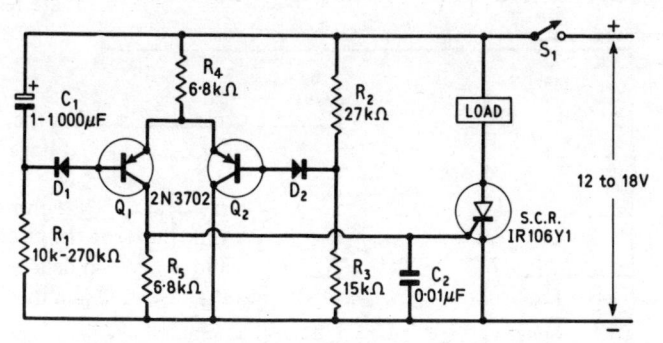

Figure 4.4. Precision delayed turn-on d.c. switch, giving delays up to several minutes. D_1 and D_2 = general purpose silicon diodes.

The *Figure 4.5* circuit uses a four-transistor differential detector, with each side of the detector super-alpha connected. This circuit can give precision delays up to one hour. In both circuits the time-delay is approximately equal to the $R_1 - C_1$ product.

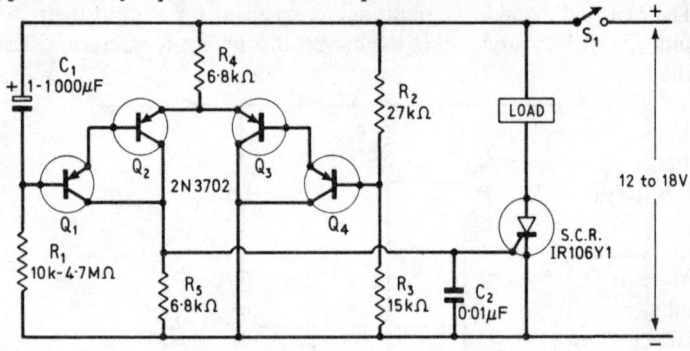

Figure 4.5. Precision long-period delayed turn-on d.c. switch, giving delays up to one hour.

D.C. auto-turn-off projects

Auto-turn-off circuits are designs in which power is applied to the load as soon as an operating switch is closed, but the power is then automatically removed again after a preset delay. A number of alternative techniques can be used to achieve this type of operation.

Figure 4.6 shows the practical circuit of an auto-turn-off d.c. switch using two s.c.r.s. In this design power is permanently applied to the circuit, and SCR_1 is normally off and SCR_2 is on. Under this condition

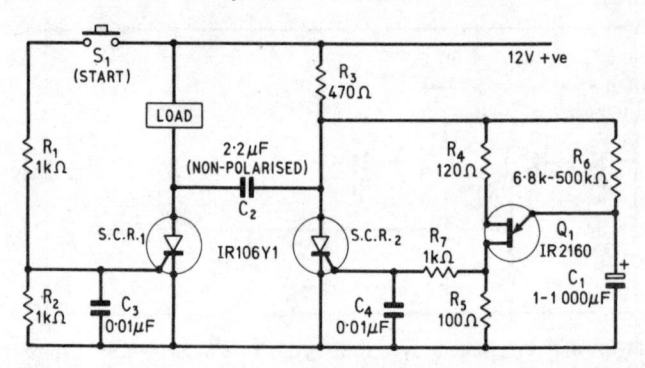

Figure 4.6. Auto-turn-off d.c. switch, giving delays up to 15 minutes.

zero power is developed across the load, near-zero voltage is developed across the u.j.t. timer network, and C_2 is fully charged with its SCR_1-anode side positive.

Power is applied to the load by briefly closing push-button switch S_1, thus driving SCR_1 on. As SCR_1 goes on its anode drops sharply to near-zero volts, and thus forces the anode of SCR_2 to swing momentarily negative via the charge of C_2. As SCR_2 anode swings negative its anode current falls to zero, and it turns off.

Once SCR_2 has turned off its anode voltage rises close to that of the positive supply line, and power is applied to the u.j.t. timer network. Simultaneously, C_2 charges rapidly via R_3, its SCR_2-anode end becoming positive. As soon as power is applied to the u.j.t. timer network C_1 starts to charge exponentially via R_6, and after a preset period the u.j.t. fires and turns SCR_2 back on. As SCR_2 goes on it forces the anode of SCR_1 sharply negative via the charge of C_2, so SCR_1 turns off and removes power from the load. C_2 then recharges rapidly in the reverse direction via the load, its SCR_1-anode end becoming positive again. The operating cycle of the circuit is then complete.

The only snag with the *Figure 4.6* circuit is that it needs to be permanently connected to a power supply, and thus consumes a constant standby current of about 20 mA (via R_3). This snag can be overcome by using the s.c.r.-plus-relay circuit shown in *Figure 4.7*.

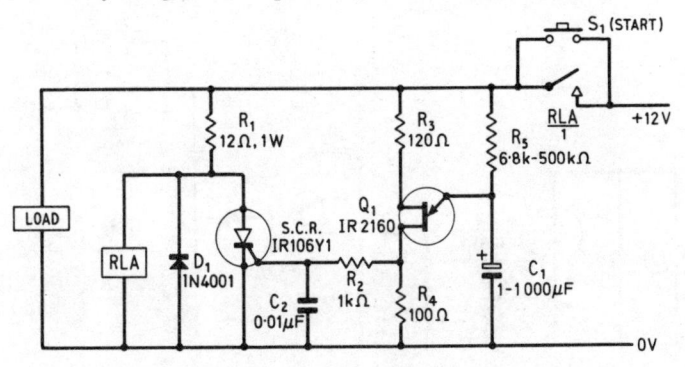

Figure 4.7. SCR relay auto-turn-off d.c. switch, giving delays up to 15 minutes.
RLA = 12 V relay with coil resistance in the range 120 Ω to 1000 Ω: one set of N/O contacts.

Here, power is connected to the load via the contacts of relay RLA, and the s.c.r. and u.j.t. circuitry are used to control the relay switching action. The relay has a coil resistance that is large relative to R_1.

Normally, the relay is off, so contact $RLA/1$ is open and the circuit consumes zero power. Circuit action is initiated by momentarily closing push-button switch S_1, thus applying power to the load and the electronic circuitry. When S_1 is first closed the s.c.r. is off. Since the coil resistance of RLA is large relative to R_1, therefore, almost the full supply voltage appears across the relay, which turns on. As RLA goes on contacts $RLA/1$ close and cause the relay to self-latch, thus maintaining power to the circuit once S_1 is released.

As soon as power is applied to the circuit C_1 starts to charge up via R_5. Eventually, after a preset delay, C_1 reaches the firing voltage of the u.j.t., which then fires and triggers the s.c.r. on. As the s.c.r. turns on its anode voltage falls sharply to near-zero, thus effectively removing the supply from the relay and causing it to turn off. As the relay turns off contacts $RLA/1$ open and remove all power from the load and the electronic circuitry. The circuit action is then complete.

The relay used in this circuit can be any 12 V type with a coil resistance in the range of 120 ohms to 1 000 ohms, and with one or more sets of suitably rated N/O contacts (to supply the required load current).

Note that the u.j.t. timer used in the *Figure 4.7* circuit is identical to that used in *Figure 4.1*, and that the circuit can be used to give delays up to 15 minutes. The transistor timing circuits of *Figures 4.2* to *4.5* can be used to replace the u.j.t. timer, if required, and *Figures 4.8* to *4.10* show how the *Figures 4.2* to *4.4* circuits can be so con-

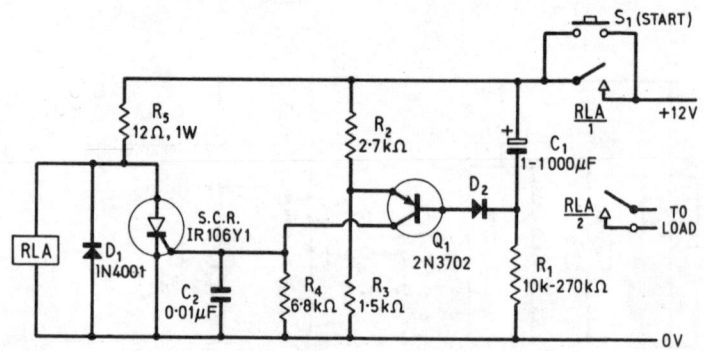

Figure 4.8. SCR + relay auto-turn-off d.c. switch, giving delays up to several minutes.
  RLA = 12 V relay with coil resistance in the range 120 Ω to 1000 Ω: 2 sets of N/O contacts.
  D_1 = general-purpose silicon diode.

nected. Note in these three circuits that the relay has two sets of N/O contacts, and that the $RLA/2$ set are used to control the load. If the load is to be operated from the same 12 V supply as the timer circuit, however, contacts $RLA/2$ can be eliminated and the load can be wired directly across the time-delay network, as in the case of *Figure 4.7*.

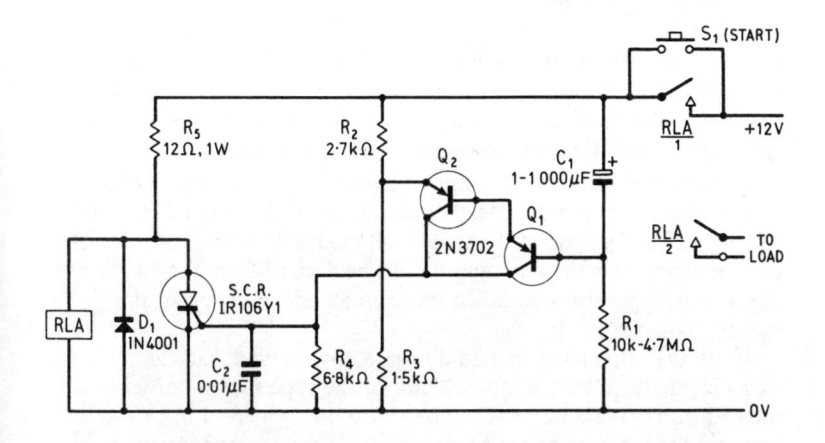

Figure 4.9. Long-period SCR + relay auto-turn-off d.c. switch, giving relays up to one hour.
RLA = 12 V relay with a coil resistance in the range 120 Ω to 1000 Ω: 2 sets of N/O contacts.

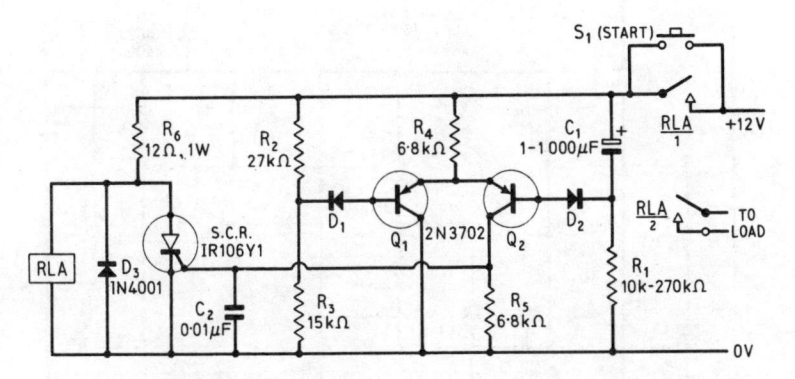

Figure 4.10. Precision SCR + relay auto-turn-off d.c. switch, giving delays up to several minutes.
RLA = 12 V relay with a coil resistance in the range 120 Ω to 1000 Ω: 2 sets of N/O contacts.
D_1 and D_2 = general purpose silicon diodes.

A.C. time-delay projects

The most satisfactory method of achieving a.c. time-delay switching is to use a triac as the a.c. switching element, and to use a d.c. network to provide its time-delayed gating control. The actual d.c. gate drive signal can be of either the smoothed or the pulsed type.

If a smoothed d.c. signal is used for gate control the drive network must be designed to pass a steady gate current of at least 120 mA, to ensure stable triggering of the triac in all modes. Because of these high drive requirements the d.c. supply must be derived from the a.c. line via a relatively expensive step-down transformer and a rectifying and smoothing network.

If, on the other hand, pulsed d.c. gate drive is used, and the pulsed signal is generated by a u.j.t. oscillator of the type shown in *Figure 2.5*, the gate current requirement can be reduced to a mere 4 mA or so. In this case the d.c. supply can be derived from the a.c. line via an inexpensive dropper resistor and a rectifier and smoothing network. Thus, the pulsed d.c. system is the most economic method of control.

Figure 4.11 shows the practical circuit of a delayed turn-on line switch using pulsed d.c. gate triggering. Q_3 is the u.j.t. pulse generator,

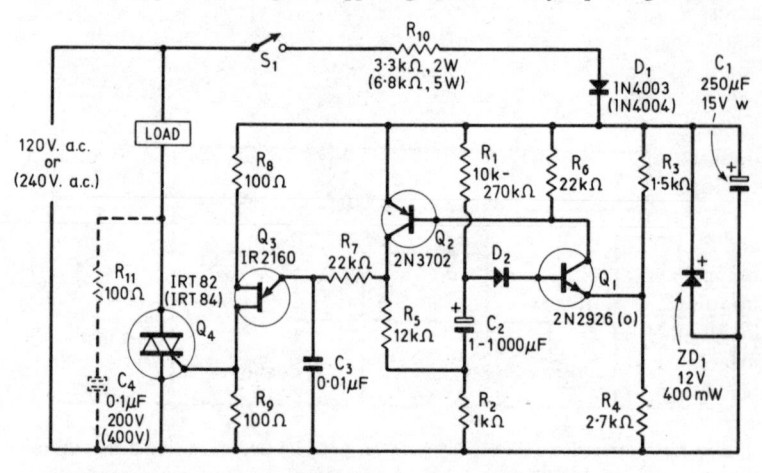

Figure 4.11. Delayed turn-on line switch, giving delays up to several minutes. D_1 = general purpose silicon diode.

which is controlled by the $Q_1 - Q_2$ time-delayed switching circuit. The d.c. supply of the circuit is derived from the a.c. line via $R_{10} - D_1$ and zener diode ZD_1 and smoothing capacitor C_1. The time-delay section of the circuit is roughly similar to the transistor time-delay circuits already described, and operation is as follows.

15 TIME-DELAY PROJECTS 63

$R_1 - C_2$ and R_2 are wired across the d.c. supply as a time-controlled potential divider that applies a variable voltage to Q_1 base via D_2, and R_3 and R_4 are wired across the d.c. supply so that they generate a fixed reference voltage that is applied to Q_1 emitter. Q_1 collector current feeds directly into Q_2 base, and Q_2 collector controls the u.j.t. oscillator via R_7. Part of the Q_2 collector current is fed back to the Q_1 timer network via R_5, and forms a regenerative feedback path between Q_1 and Q_2.

When power is first applied to the circuit, by closing S_1, C_2 is fully discharged and acts like a short circuit. Under this condition the Q_1 base-emitter junction is reverse biased, and Q_1 is cut off. Zero base current thus flows in Q_2, which is also cut off. Since Q_2 is cut off zero current flows in R_7, so the u.j.t. oscillator is inoperative and the triac is off. Zero power is thus developed in the load at this time.

C_2 starts to charge exponentially via R_1 and R_2 as soon as S_1 is closed, and the C_2 voltage slowly rises. Eventually, after a preset delay, the C_2 voltage rises to such a level that the Q_1 base-emitter junction starts to become forward biased, and Q_1 begins to conduct. The resulting Q_1 collector current is fed into the base of Q_2, where it is amplified and fed back to R_2 via R_5, causing a rising voltage to develop across R_2. This rising voltage is coupled to Q_1 base via C_2 and D_2, and causes the Q_1 collector current to increase further. A regenerative action thus takes place as soon as the Q_1 base-emitter junction becomes forward biased, and Q_2 is driven sharply to saturation. Under this condition current flows to the u.j.t. circuit via R_7, and the u.j.t. circuit oscillates and drives the triac on, and full power is thus developed in the load.

R_1 in this circuit can have any value in the range 10 k to 270 k ohms, and C_2 can have any value up to 1 000 μF. The time-delay of the circuit is approximately equal to the $R_1 - C_2$ product, and maximum delays up to several minutes are possible. Longer delays, up to one hour, can be obtained by increasing the R_1 value (up to a maximum of 4·7 megohms) and using a super-alpha connected pair of transistors in place of Q_1, as shown in the long-period delayed turn-on switch circuit of *Figure 4.12*.

Figure 4.13 shows how the *Figure 4.11* circuit can be modified so that it acts as an auto-turn-off line switch. Note that the Q_1 time-delay circuit is inverted, and that the Q_2 switching transistor is arranged so that it shunts the u.j.t. oscillator network. Circuit operation is as follows.

When power is first applied to the circuit, by closing S_1, C_2 is fully discharged and acts like a short circuit. Under this condition the Q_1 base-emitter junction is reverse biased, and Q_1 is cut off. Zero base current thus flows in Q_2, which is also cut off. Since Q_2 is cut off,

64 15 TIME-DELAY PROJECTS

current flows to the u.j.t.s timing capacitor via $R_7 - R_5$ and R_2, and the u.j.t. circuit oscillates and drives the triac on. Full power is thus developed in the load at this time.

C_2 starts to charge exponentially via R_1 and R_2 as soon as S_1 is closed, and the C_2 charge slowly rises. Eventually, after a preset delay, the C_2 voltage rises to such a value that the Q_1 base-emitter junction

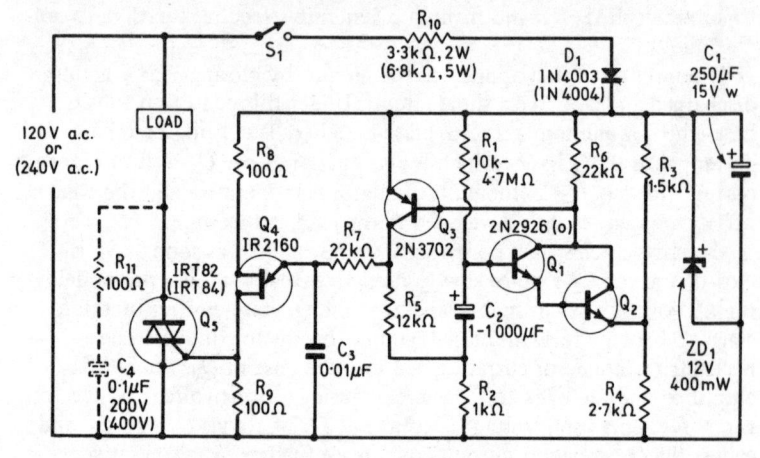

Figure 4.12. Long-period delayed turn-on line switch, giving delays up to one hour.

starts to become forward biased, and Q_1 begins to conduct. The resulting Q_1 collector current is fed into the base of Q_2, where it is amplified and fed back to R_2 via R_5, causing a falling voltage to develop across R_2. This voltage is coupled to Q_1 base via C_2 and D_2, and causes the Q_1 collector current to increase further. A regenerative action thus takes place as soon as the Q_1 base-emitter junction becomes forward biased, and Q_2 is driven sharply to saturation. Under this condition zero current flows to the u.j.t. circuit via R_7, so the u.j.t. stops oscillating and the triac turns off. Zero power is thus developed in the load under this condition.

R_1 in the *Figure 4.13* circuit can have any value in the range 10 k to 270 ohms, and the circuit can give maximum delays up to several minutes. Longer delays, up to one hour, can be obtained by increasing the R_1 value and using a super-alpha connected pair of transistors in place of Q_1, as shown in the long-period auto-turn-off line switch circuit of *Figure 4.14*. This diagram also shows how the circuit can be modified for push-button activation. This action is achieved by wiring a line-driven relay across the a.c. load, and shunting the relays N/O contacts with push-button switch S_1 and then wiring the combination

15 TIME-DELAY PROJECTS 65

in series with R_{10}. Note that an $R_{11} - C_4$ 'snubber' network is permanently wired across the triac in this circuit, to prevent circuit triggering via unwanted line transients.

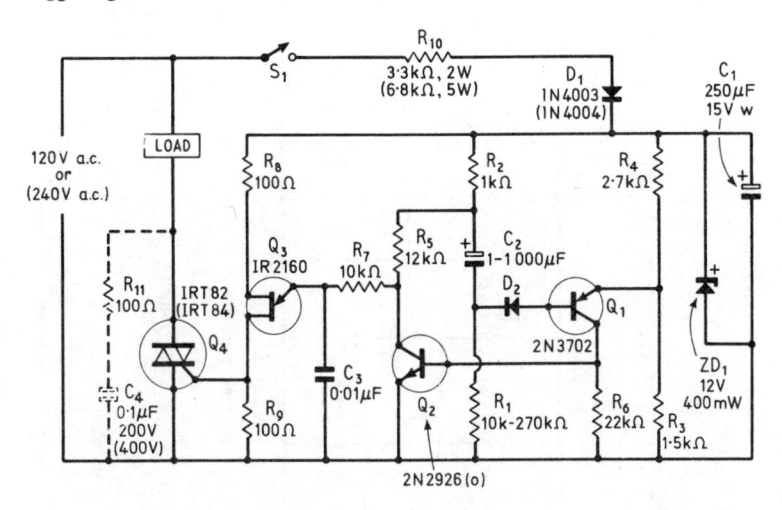

Figure 4.13. Auto-turn-off line switch, giving delays up to several minutes. D_2 = general purpose silicon diode.

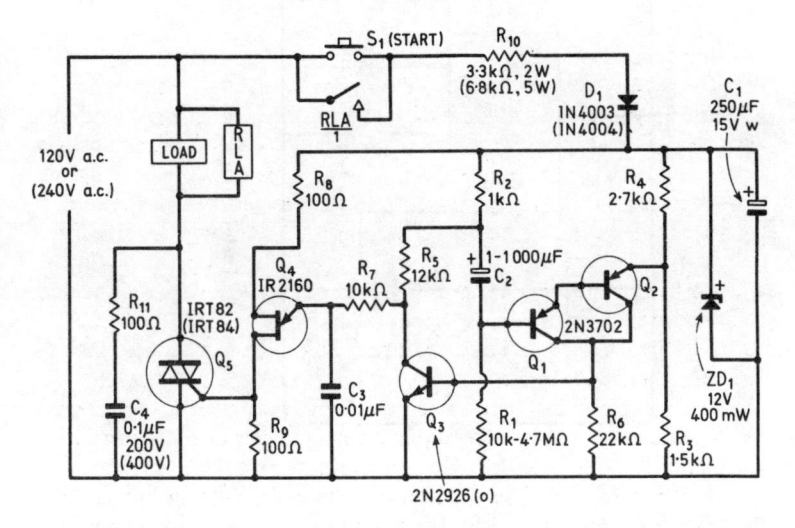

Figure 4.14. Long-period push-button activated auto-turn-off line switch, giving delays up to one hour.
  RLA = line-powered relay with one or more sets of N/O contacts.

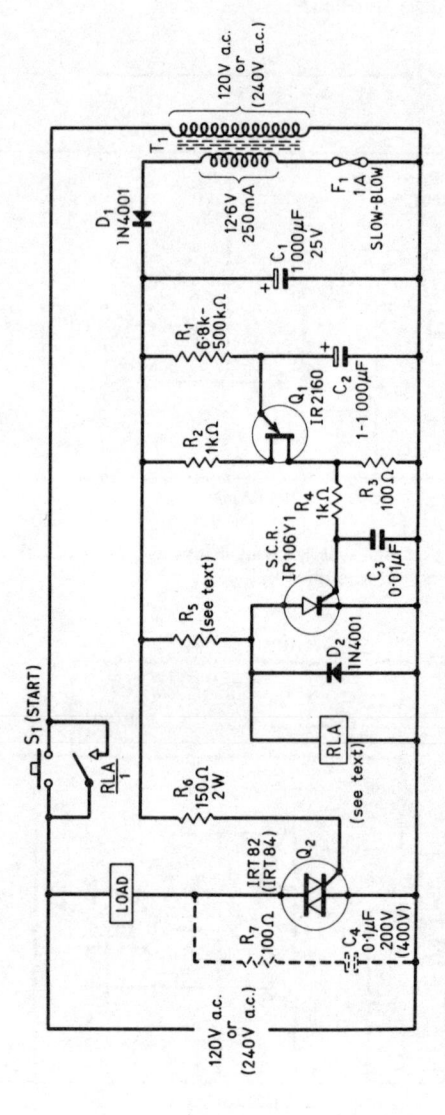

Figure 4.15. Push-button activated auto-turn-off line switch, giving delays up to 15 minutes.

15 TIME-DELAY PROJECTS 67

Finally, *Figure 4.15* shows the circuit of another push-button activated auto-turn-off line switch. This circuit uses steady d.c. gate triggering of the triac, and uses a low-voltage d.c. relay to control the gate action. To meet the heavy d.c. low-voltage requirements of the circuit the d.c. supply is derived from the a.c. line via a 12·6 V step-down transformer. Circuit operation is as follows.

Normally, relay *RLA* is off and contacts *RLA*/1 are open, so zero power is developed in the low-voltage d.c. circuit. The triac is thus off and zero power is developed in the load under this condition. Circuit operation is initiated by briefly closing S_1, thus applying power to the d.c. circuit. When S_1 is first closed the s.c.r. is off. The coil resistance of *RLA* is large relative to R_5, so a large voltage appears across *RLA* and the relay immediately turns on. As *RLA* goes on contacts *RLA*/1 close and shunt S_1, thus causing the relay to self-latch. Simultaneously, d.c. gate drive is applied to the triac via R_6, and the triac turns on and applies full power to the load.

As soon as S_1 is closed C_2 starts to charge up via R_1. Eventually, after a preset delay, C_2 reaches the firing voltage of the u.j.t., which then fires and triggers the s.c.r. on. As the s.c.r. turns on its anode voltage falls sharply to near zero, thus effectively removing the supply from the relay and causing it to turn off. As the relay turn off contacts *RLA*/1 open and disconnect the d.c. supply from the circuit, and the triac turns off. The circuit action is then complete.

The relay used in this circuit can be any 12 V type with a coil resistance in the range of 120 ohms to 1 000 ohms, and with one or more sets of N/O contacts. R_5 must have a value roughly half that of the relay coil, so that roughly 12 V is developed across the relay when the s.c.r. is off; e.g., if the relay has a 470 ohm coil, R_5 needs a value of 220 or 270 ohms.

Note that the u.j.t. timer used in the *Figure 4.15* circuit is identical to that used in *Figure 4.7*, and that delays up to 15 minutes are possible. The transistor timing circuits of *Figures 4.8* to *4.10* can be used to replace the u.j.t. timer, if required, and in this case delays up to one hour are possible, providing that good-quality and low-leakage timing capacitors are used.

CHAPTER 5

25 LAMP-CONTROL PROJECTS

S.C.R.s and triacs can be used in a wide range of a.c. and d.c. incandescent lamp control projects. Twenty-five such projects are described in this chapter, and include automatic turn-on and turn-off circuits, time-delayed circuits, lamp flashers and lamp chasers, and a variety of lamp dimmers and special control circuits.

Automatic (light activated) a.c. lamp projects

Figure 5.1 shows a simple light-activated circuit that turns an a.c. lamp on automatically when it gets dark, and turns it off again automatically when it gets light. Circuit operation is fairly simple. The lamp is controlled by the triac, which has pulsed d.c. gate triggering applied via u.j.t. oscillator Q_2. The u.j.t.s d.c. supply is derived from the a.c. line via $R_5 - D_1 - ZD_1$ and C_1, and this supply also provides power to the light-controlled potential divider formed by R_1 and *LDR*. This potential divider provides emitter drive to the u.j.t. oscillator via R_2 and D_2, and R_1 is adjusted so that the voltage at the $R_1 - LDR$ junction is slightly in excess of the u.j.t.s peak-point or firing voltage (typically 0·7 times the d.c. supply voltage) at the required turn-on light level.

Thus, under bright conditions the *LDR* presents a low resistance, so the voltage on the $R_1 - LDR$ junction is below the u.j.t.s peak-point voltage. Under this condition, therefore, C_2 is unable to charge to the firing voltage of the u.j.t., so the u.j.t. oscillator is inoperative and the triac and lamp are off. Under dark conditions, on the other hand, the *LDR* resistance is high and the voltage on the $R_1 - LDR$ junction is above the u.j.t.s firing voltage, so under this condition C_1 is able to

25 LAMP-CONTROL PROJECTS 69

charge to the u.j.t.s firing voltage and the u.j.t. circuit oscillates normally and drives the triac and the lamp on. The circuit thus gives completely automatic on-off control of the lamp.

A minor disadvantage of the *Figure 5.1* circuit is that it generates a certain amount of lamp flicker when the ambient light level is very close to the preset turn-on light level. This flickering may be objectionable if

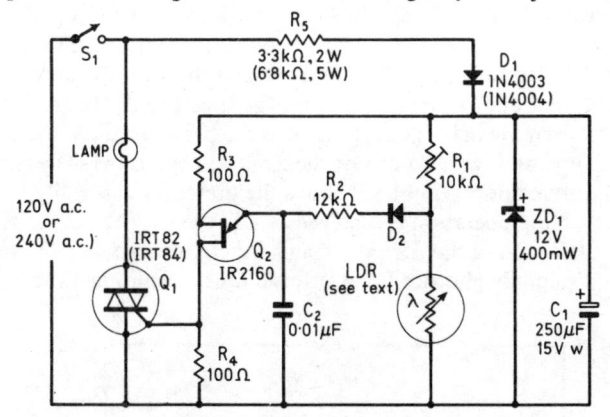

Figure 5.1. Simple light-activated a.c. lamp, for outdoor use. D_2 = general purpose silicon diode.

the circuit is used for indoor light control, and for this reason it is recommended that the design be used to control outdoor lights only.

Figure 5.2 shows an improved light-activated lamp control circuit that is suitable for indoor use, and which turns the lamp on and off with no sign of flickering. The circuit again uses a d.c. pulse-triggered triac to control the lamp, but in this case the u.j.t. oscillator is controlled by a precision light-activated switch formed by the $Q_3 - Q_4$ and *LDR* network.

$R_1 - LDR$ and R_2 are wired as a light-controlled potential divider that applies a variable voltage to Q_4 base, and $R_3 - R_4$ are wired as a potential divider that applies a fixed reference voltage to Q_4 emitter. R_1 is adjusted so that the Q_4 base voltage is slightly above that of the emitter at the required turn-on light level.

Thus, under conditions of bright illumination the *LDR* presents a low resistance and the Q_4 base voltage is below that of the emitter, so Q_4 is cut off. Since Q_4 is cut off zero base current is fed to Q_3, which is also cut off, and since Q_3 is cut off zero current flows to the u.j.t.s emitter via R_7, and the u.j.t. oscillator is inoperative. Thus, the triac and the lamp are off under this condition.

As the ambient light level falls the *LDR* resistance increases and the voltage on Q_4 base rises. Eventually, when the light intensity falls to

70 25 LAMP-CONTROL PROJECTS

the preset turn-on level, the Q_4 base voltage rises to such a level that the Q_4 base-emitter junction becomes forward biased and the transistor starts to conduct. The resulting collector current of Q_4 feeds directly into the base of Q_3, where it is amplified, and the resulting collector current of Q_3 is fed back to Q_4 base via $R_5 - R_2$ and the LDR and causes the Q_4 base voltage to increase further. Q_4 thus turns on harder and increases the drive to Q_3, so a regenerative action takes place in which Q_3 suddenly switches to saturation. Under this condition emitter current flows to the u.j.t. via R_7 and Q_3, and the u.j.t. oscillates and drives the triac and the lamp on. Thus, the triac and the lamp turn on abruptly when the light intensity falls to the preset turn-on level, and the lamp operates with no sign of flicker. The lamp turns off again when the LDR illumination rises back above the preset level, so fully automatic lamp operation is achieved.

The LDR used in the *Figure 5.1* and *5.2* circuits can be any small cadmium sulphide photocell that offers a resistance in the range 1 000

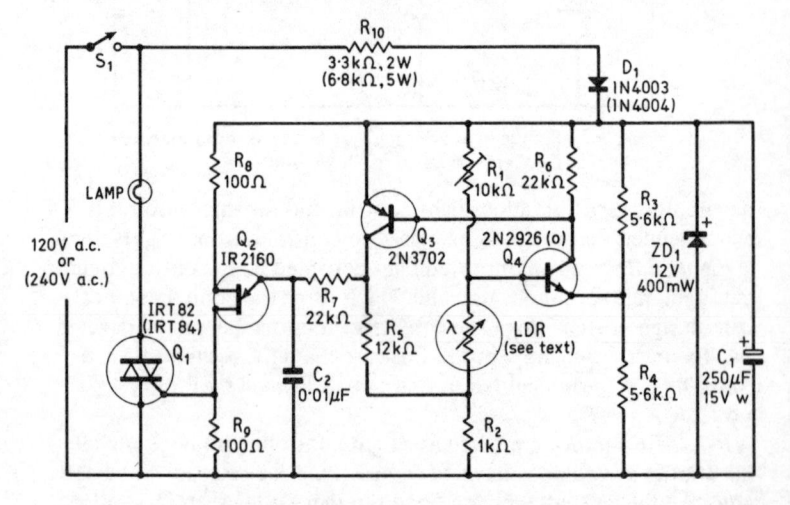

Figure 5.2. Improved light-activated a.c. lamp, for indoor use.

to 10 000 ohms at the required turn-on light level. This cell should be mounted at least seven feet above the ground, so that it is not influenced by shadows thrown by passing people or objects, and should be positioned so that it responds to the mean light level of the room or area in which the controlled lamps are situated. The cell should not be placed in deep shadow or in a position where it is excessively illuminated by a single point of illumination.

When setting up the circuits by adjusting R_1, it is important to note that, for satisfactory operation, the illumination cast on to the LDR face by the lamp must be lower than that cast by the natural illumination at the required preset turn-on level. If this point is not observed a photoelectric positive feedback loop will be created between the lamp and the *LDR,* and the circuit will then function as a lamp flasher or oscillator.

Time-delayed lamp-control projects

All of the time-delay circuits shown in Chapter 4 can be used to control incandescent lamps, the final choice of circuit being a matter of individual preference. *Figures 5.3* to *5.6* show the author's personal choice of delayed turn-on and auto-turn-off d.c. and a.c. lamp-control circuits.

Figure 5.3 shows a delayed turn-on d.c. lamp circuit that can give delays up to 15 minutes. The lamp in this circuit can be any 12 V type

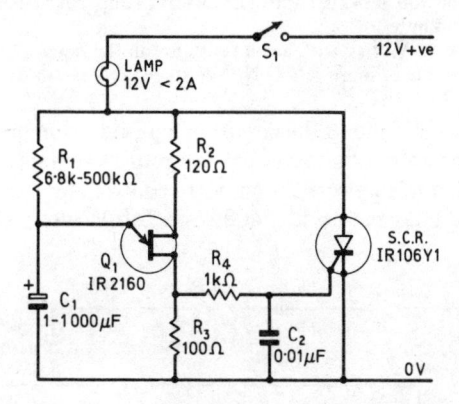

Figure 5.3. Delayed turn-on d.c. lamp, giving delays up to 15 minutes.

that draws a current less than 2 A. The circuit is developed from *Figure 4.1,* and its operation is described in Chapter 4.

Figure 5.4 shows a push-button activated auto-turn-off d.c. lamp circuit that gives delays up to 15 minutes. The lamp in this circuit is controlled by relay contacts *RLA*/1, and the lamp can thus be any 12 V type with a current rating that can be handled by the *RLA*/1 contacts. The relay can be any 12 V type with a coil resistance in the range 120 to 1 000 ohms, and with one or more sets of suitably rated N/O contacts. The *Figure 5.4* circuit is identical to that of *Figure 4.7.*

Figure 5.5 shows a delayed turn-on a.c. lamp circuit that gives delays up to several minutes. Using the specified triacs the circuit can be used to control incandescent lamp loads up to 960 W on 120 V a.c. lines, or

72 25 LAMP-CONTROL PROJECTS

up to 2 400 W on 240 V lines. The circuit is almost identical to that of *Figure 4.11*, and a full description is given in Chapter 4.

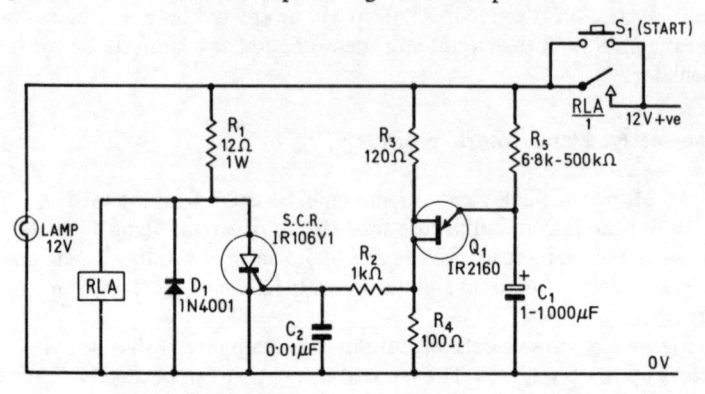

Figure 5.4. Push-button activated auto-turn-off d.c. lamp, giving delays up to 15 minutes.
RLA = 12 V relay with a coil resistance in the range 120 Ω to 1000 Ω: one or more sets of N/O contacts.

Finally, *Figure 5.6* shows the circuit of a push-button operated auto-turn-off line switch that gives delays up to 15 minutes. The circuit is identical to that of *Figure 4.5*, and it can control lamp loads up to 960 W on 120 V lines and up to 2 400 W on 240 V lines.

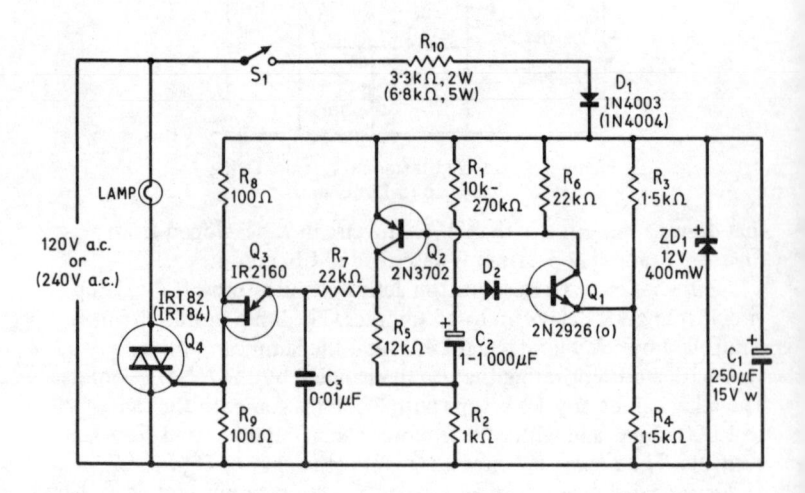

Figure 5.5. Delayed turn-on a.c. lamp, giving delays up to several minutes.
D_2 = general purpose silicon diode.

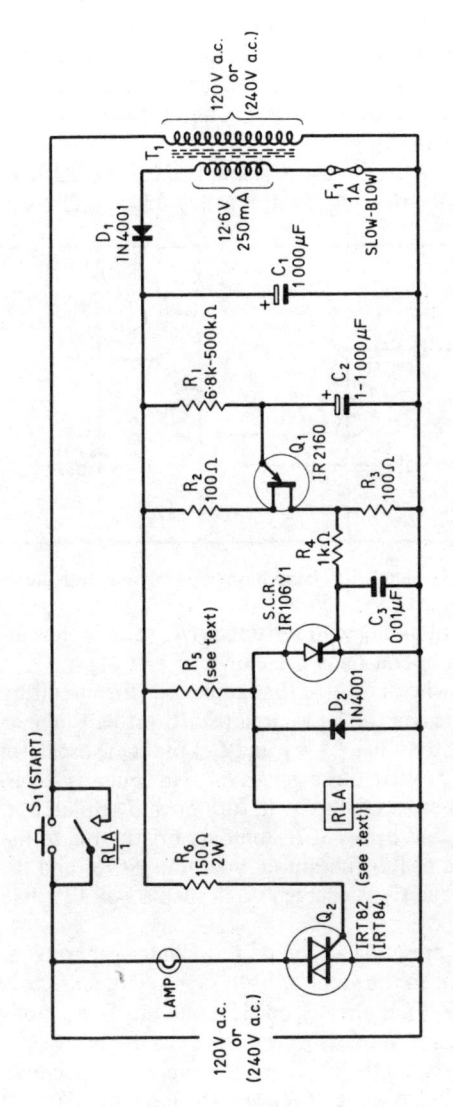

Figure 5.6. Push-button activated auto-turn-off line switch, giving delays up to 15 minutes.

D.C. lamp flasher projects

D.C. lamp flashers can be readily designed around either conventional bipolar transistors or around s.c.r.s. Transistor designs are, frankly, to be preferred on grounds of cost and efficiency. S.C.R. designs are, however, of considerable technical interest, and for this reason three such designs are shown in this section.

Figure 5.7 shows the circuit of a simple d.c. lamp flasher, in which Q_1 is wired as a free-running low-frequency u.j.t. oscillator, and SCR_2

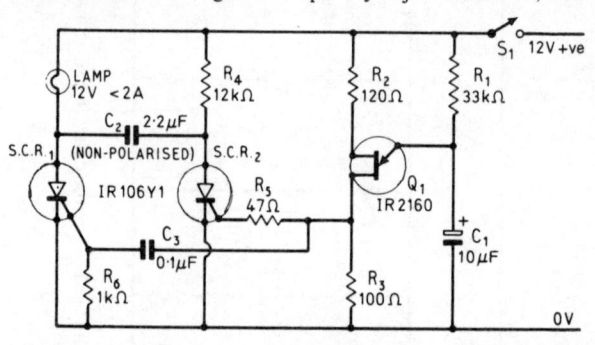

Figure 5.7. Simple d.c. lamp flasher, giving one flash per second.

has a large value of anode load resistance (R_4) and is thus unable to self-latch. Circuit operation is as follows.

Assume that, when power is first applied to the circuit by closing S_1, both s.c.r.s are off and the u.j.t. circuit starts into a timing cycle. After a preset delay (determined by R_1 and C_1) the u.j.t. oscillator fires and generates a large positive pulse across R_3. This pulse is fed to the gate of SCR_1 via C_3 and causes the s.c.r. to turn on and self-latch; the pulse is also fed to SCR_2, but drives SCR_2 on only briefly due to the high value of R_4. At the end of this first pulse, therefore, SCR_1 and the lamp are on, SCR_2 is off, and C_2 is charged up in such a way that its R_4 end is positive.

After another preset delay the u.j.t. oscillator generates another pulse, which is fed to the gates of both s.c.r.s. The pulse to SCR_1 has no effect, since SCR_1 is already on. The pulse to SCR_2, however, drives SCR_2 briefly on, and as SCR_2 goes on it pulls the positive end of C_2 to ground and thus causes the C_2 charge to apply a reverse bias to the anode of SCR_1, thus turning SCR_1 and the lamp off. SCR_2 again turns off at the end of the pulse, due to its lack of holding current. At the end of the second pulse, therefore, both s.c.r.s are again off, and the entire switching process is ready to start over again on the arrival of the third pulse from the u.j.t. oscillator.

25 LAMP-CONTROL PROJECTS 75

The circuit thus acts as a lamp flasher in which the lamp is sequentially turned on and off for equal periods at a rate equal to half that of the u.j.t. oscillator. The flashing rate is roughly equal to twice the $R_1 - C_1$ product, and is approximately one flash per second with the component values shown. The rate can be increased by reducing the R_1 value, down to a minimum of 6·8 k ohms, or can be decreased by raising the R_1 value, up to a maximum of 470 k ohms.

Figure 5.8 shows how the *Figure 5.7* circuit can be modified so that it turns on automatically at dusk and off again at dawn. To understand

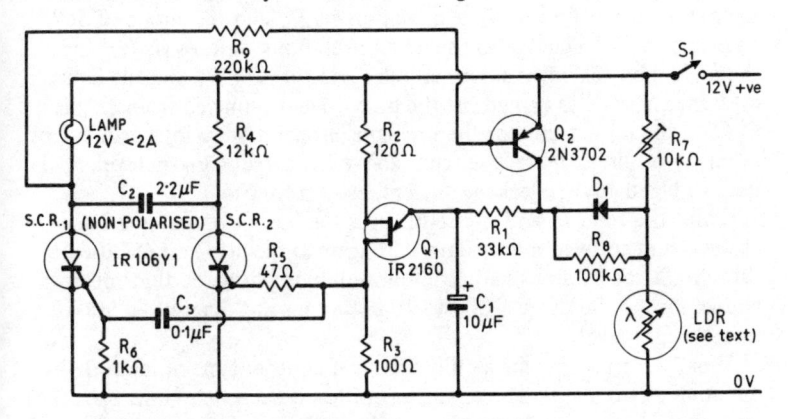

Figure 5.8. Automatic (light activated) d.c. lamp flasher: turns on at dusk and off at dawn.
D_1 = general purpose silicon diode.

the initial explanation of circuit operation, assume that $Q_2 - R_9$ and R_8 are removed from the circuit. In this case it can be seen that the circuit is similar to that of *Figure 5.7*, except that the u.j.t.s main timing resistor, R_1, is taken to the $R_7 - LDR$ junction via D_1. R_7 and the LDR are wired as a light-dependent potential divider, and in use R_7 is adjusted so that the voltage on the $R_7 - LDR$ junction is slightly in excess of the u.j.t.s peak-point voltage at the required 'dusk' turn-on light level.

Thus, when the light intensity is above the preset turn-on level the LDR presents a low resistance and the voltage at the $R_7 - LDR$ junction is below the u.j.t.s peak-point voltage. Under this condition C_1 is unable to charge to the firing voltage of the u.j.t., so the u.j.t. oscillator and the lamp flasher are inoperative

When, on the other hand, the light intensity is below the preset level, the LDR resistance is high and the voltage on the $R_7 - LDR$ junction is above the u.j.t.s peak-point voltage. Under this condition, therefore, C_1 is able to charge to the u.j.t.s firing voltage via R_1 and R_7, and the u.j.t.

circuit oscillates and causes the unit to act as a lamp flasher. Thus, the flasher is turned on and off automatically at dusk and dawn.

A major snag of the circuit described above is that, if SCR_1 is on at the moment that the u.j.t. stops oscillating at dawn, SCR_1 will lock on and illuminate the lamp throughout the entire day. In the full circuit of *Figure 5.8* this snag is overcome by $Q_2 - R_9$ and R_8. Q_2 is wired between R_1 and the positive supply line, and has its base bias derived from the anode of SCR_1 via R_9.

Consequently, if SCR_1 is on when the $R_7 - LDR$ voltage falls below the preset turn-off level, Q_2 is driven on via R_9 and the anode of SCR_1 and enables C_1 to charge to the u.j.t.s peak-point voltage via R_1, and the u.j.t. then generates a further pulse and turns SCR_1 and the lamp off. Once SCR_1 has turned off the base drive is removed from Q_2 via R_9, and the u.j.t. oscillator then remains inoperative so long as the light intensity on the LDR face remains above the preset turn-on level. R_8 is used to bleed any Q_2 leakage current away from C_1.

Thus, the full *Figure 5.8* circuit turns the flasher on automatically at dusk, and turns it and the lamp off again at dawn. The LDR used in this circuit can be any small cadmium sulphide photocell that presents a resistance in the range 1 000 to 10 000 ohms at the required 'dusk' turn-on light level.

Finally, *Figure 5.9* shows the circuit of another type of lamp flasher, employing two lamps. The circuit operation is such that lamp 1 is off when lamp 2 is on, and vice versa, and is as follows.

When power is initially applied to the circuit by closing S_1 all capacitors are discharged. C_5 thus acts as a short circuit at this instant,

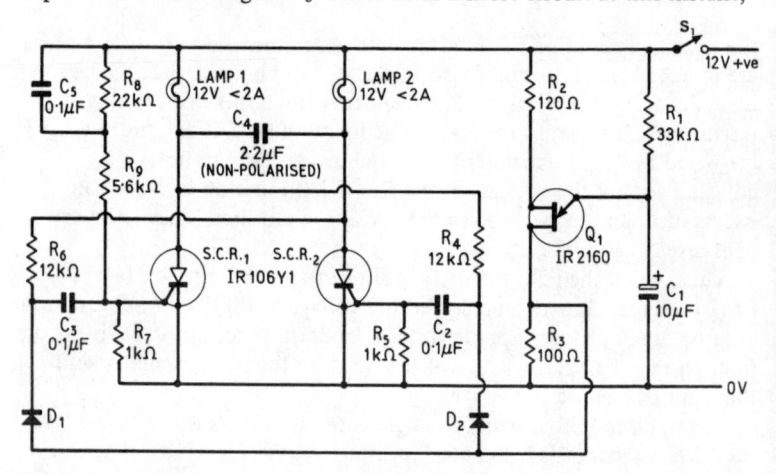

Figure 5.9. Twin-lamp d.c. flasher.
D_1 and D_2 = general-purpose silicon diodes.

so a brief pulse of gate drive is applied to SCR_1 via C_5 and R_9, and SCR_1 and lamp 1 turn on. C_5 charges rapidly when S_1 is first closed, and then acts like an open circuit, in which case R_8 prevents further gate drive from reaching SCR_1 via the positive supply line. Thus, SCR_1 and lamp 1 turn on as soon as S_1 is closed, and C_4 charges up in such a way that its lamp 2 end is positive. Simultaneously, C_1 starts to charge up via R_1 in the u.j.t. oscillator circuit.

After a preset delay C_1 reaches the peak-point voltage of Q_1 and the u.j.t. fires and generates a large positive pulse across R_3. At this moment D_1 is reverse biased via R_6 and lamp 2, and D_2 is forward biased, so the pulse is coupled to the gate of SCR_2 only, and SCR_2 and lamp 2 turn on. As SCR_2 turns on it pulls the positive end of C_4 to ground, and the C_4 charge then applies reverse bias to the anode of SCR_1 and causes SCR_1 and lamp 1 to turn off. As SCR_2 turns on it removes the reverse bias from D_1, and as SCR_1 turns off it applies reverse bias to D_2. Consequently, when the u.j.t. completes its next timing cycle and generates a trigger pulse the pulse is applied to SCR_1 only, so SCR_1 and lamp 1 turn on again and cause C_4 to turn SCR_2 back off.

The circuit thus acts as a true bistable, and the two lamps turn on and off repetitively in opposition to one another at a rate determined by R_1 and C_1. With the component values shown the operating rate is roughly one flash per second, as in the case of the *Figure 5.7* circuit.

A.C. lamp flasher projects

A.C. lamp flashers can be used as visual alarms or 'attention getters', and have wide application in the world of commercial advertising. Triacs can be made to operate as highly efficient lamp flashers. They can be made to operate in either the synchronous or non-synchronous modes, and can be readily adapted to give automatic (light activated) operation.

Figure 5.10 shows a practical a.c. lamp flasher circuit which turns the lamp on and off for equal times at a cycling rate of once per second. Operation is fairly simple. The triac has d.c. pulse triggering provided by u.j.t. oscillator Q_2, and the u.j.t. is gated on and off repetitively by the $Q_3 - Q_4 - Q_5$ network. The Q_2 to Q_5 network is powered from a 12 V d.c. supply derived from the a.c. line via $R_9 - D_1 - ZD_1$ and C_1.

Q_4 and Q_5 are wired as an astable multivibrator in which D_2 and D_3 are used to protect the transistors against reverse base-emitter breakdown. The multivibrator has extra supply line smoothing provided via $R_5 - C_2$, to ensure that its timing periods are not upset by supply line ripple. The astable action is such that Q_4 alternatively saturates and turns off at a rate determined by the $R_3 - C_3$ and $R_4 - C_4$ time constants, and a square wave is developed across R_2. This square wave provides drive to R_6, the u.j.t.s main timing resistor, via emitter follower Q_3, so the Q_3-emitter end of R_6 is effectively switched alternatively between the positive and negative rails of the d.c. supply.

Thus, when Q_4 is saturated zero drive is applied to R_6, and the u.j.t. is inoperative and the triac and lamp are off. When Q_4 is cut off the emitter of Q_3 is effectively shorted to the positive low voltage d.c. rail,

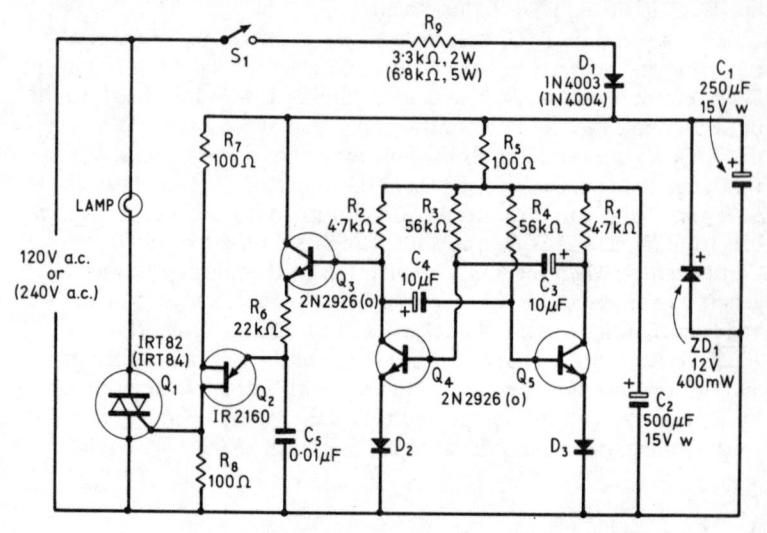

Figure 5.10. A.C. lamp flasher, giving one flash per second. D_2 and D_3 = general-purpose silicon diodes.

so drive is applied to R_6 and the u.j.t. oscillates and drives the triac and the lamp on. The lamp thus turns on and off repetitively, and the circuit acts as an a.c. lamp flasher.

Figure 5.11 shows how the above circuit can be modified so that it turns on automatically at dusk and turns off again at dawn. The circuit is similar to the above, except that light-dependent potential divider $R_{10} - LDR$ is wired in series with Q_3 emitter, and that R_6 is coupled to the $R_{10} - LDR$ junction via D_4. A low-frequency square wave is thus applied to R_6 via D_4, and has an amplitude dictated by the ambient light level. In use R_{10} is adjusted so that the peak amplitude of the square wave is slightly in excess of the u.j.t.s peak-point voltage at the required 'dusk' turn-on light level.

Thus, under bright conditions the *LDR* presents a low resistance, so the peak square wave amplitude is below the u.j.t.s peak-point voltage and the u.j.t. is inoperative. The triac and the lamp are thus permanently off under this condition. When, on the other hand, the light level is below the preset 'dusk' value the *LDR* presents a high resistance

25 LAMP-CONTROL PROJECTS   79

and the peak square wave is in excess of the u.j.t.s peak-point voltage, so the u.j.t. operates normally and the circuit acts as a lamp flasher. Fully automatic operation is thus obtained.

The *Figures 5.10* and *5.11* circuits act as excellent lamp flashers, but generate some r.f.i. when the triac turns on. If heavy lamp loads are

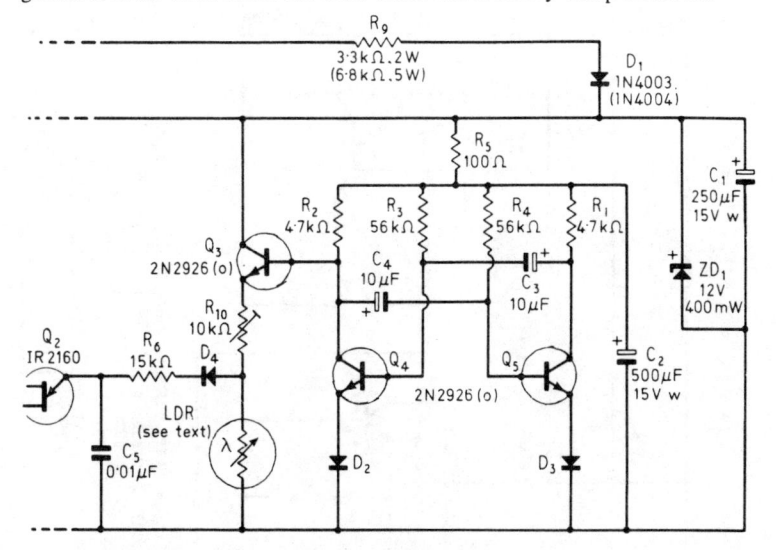

Figure 5.11. Automatic (light-activated) a.c. lamp flasher.
D_2, D_3, and D_4 = general-purpose silicon diodes.

used this r.f.i. may cause objectionable interference on a.m. radios. The r.f.i. problem can be eliminated at slight extra cost, by using the synchronous 'zero-voltage' a.c. lamp flasher circuits shown in *Figures 5.12* and *5.13*.

These two circuits are developed from the *Figure 2.10* and *Figure 2.11* synchronous line switching circuits that are fully described in Chapter 2. Looking back at these two circuits it will be remembered that they apply a trigger pulse to the triac only in the brief periods when the instantaneous line voltage is close to zero at the beginning and end of each half cycle, thus generating negligible r.f.i., and that the circuits can be turned off or inhibited by interrupting the current supply to R_5 or by driving Q_4 permanently on.

In the *Figure 5.12* flasher circuit the Q_6-Q_7 astable multivibrator is added to the basic synchronous switch and the R_5 'inhibit' resistor is wired in series with R_9 and is used as the collector load of the Q_6 multivibrator transistor. C_6 is wired between the R_5-R_9 junction and the positive low-voltage line to act as a simple filter to prevent low-

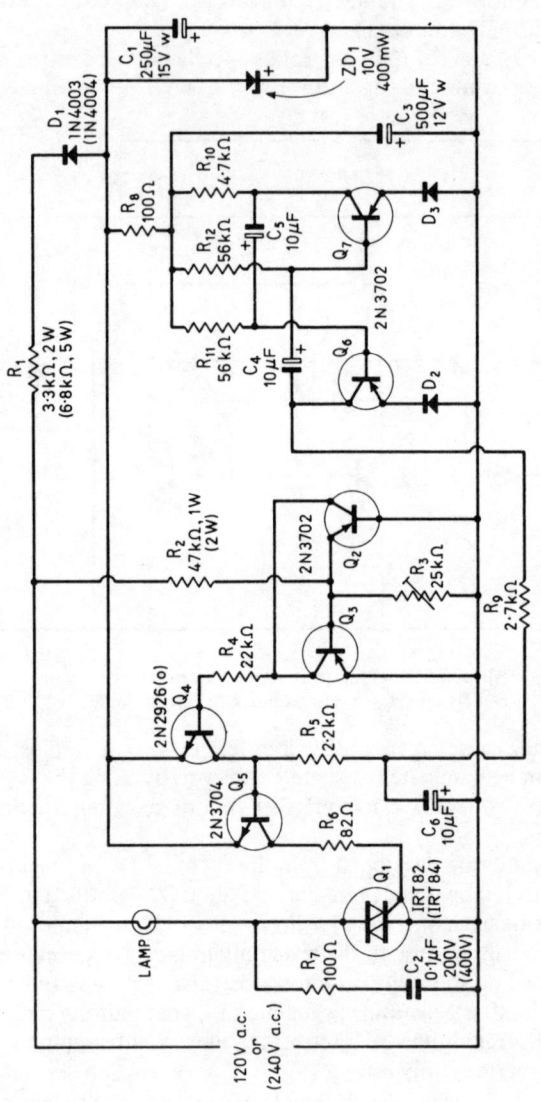

Figure 5.12. Synchronous a.c. lamp flasher, giving one flash per second. D_1 and D_3 = general-purpose silicon diodes.

voltage ripple breaking through to the astable circuit and upsetting its timing. Thus, in the periods when Q_6 is saturated heavy current flows in R_5 and R_9, and the triac and lamp are driven on synchronously. In the periods when Q_6 is cut off zero current flows in R_5 and R_9 and the circuit is inhibited, so the triac and lamp are off. The circuit thus acts as a synchronous a.c. lamp flasher, and generates negligible r.f.i.

Figure 5.13 shows how the above circuit can be modified to give automatic (light-activated) operation by the addition of $R_{13}-R_{14}-R_{15}-LDR$ and Q_8. Here, R_{13} and R_{14} are wired as a potential divider

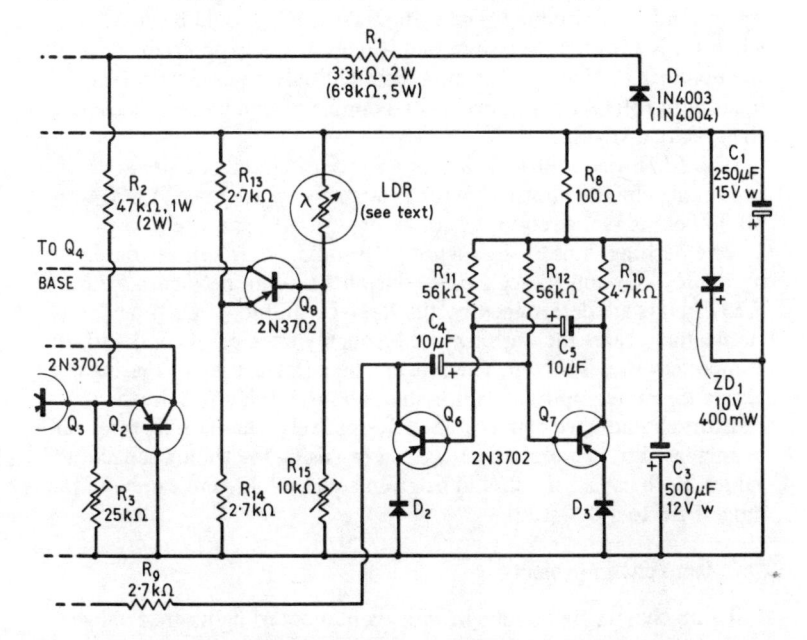

Figure 5.13. Automatic (light-activated) synchronous a.c. lamp flasher. D_2 and D_3 = general-purpose silicon diodes.

that applies a fixed negative potential to Q_8 emitter, and *LDR* and R_{15} are wired as a light-dependent potential divider that applies a variable negative voltage to Q_8 base. R_{15} is adjusted so that the Q_8 base-emitter junction is slightly forward biased at the required 'dawn' turn-off light level. The collector of Q_8 is direct-coupled to the base of Q_4.

Under dark conditions the *LDR* presents a high resistance, and the base-emitter junction of Q_8 is reverse biased and Q_8 is cut off. Q_8 thus has no effect on Q_4 under this condition, and the circuit acts as a synchronous lamp flasher in the normal way. Under bright conditions, on the other hand, the *LDR* presents a low resistance, and the base-emitter junction of Q_8 is forward biased and Q_8 is driven on. Q_8 applies base drive to Q_4 under this condition, so the circuit is inhibited and the triac and lamp are held off. The circuit thus gives fully automatic operation of the lamp flasher, and causes it to turn on at night and off during the day.

The *Figure 5.12* and *5.13* synchronous circuits are set up in the same way as already described in Chapter 2, i.e., by connecting the lamp load in place and then reducing the R_3 value until full-wave lamp operation is just obtained. The voltage across C_1 should be checked when R_3 is adjusted, to ensure that it does not fall appreciably below the nominal 10 V value. The circuit should only be used with lamp loads with total ratings in excess of a couple of hundred watts using the specified triacs.

The *LDR* used in the *Figure 5.11* and *5.13* circuits can be any cadmium sulphide photocell with a resistance in the range 1 000 to 10 000 ohms at the required turn-on or turn-off light level.

The flashing rates of the *Figure 5.10* and *5.11* circuits are determined by the $R_3 - C_3$ and $R_4 - C_4$ time constants, and in the *Figure 5.12* and *5.13* circuits are determined by the $R_{11} - C_5$ and $R_{12} - C_4$ time constants. In all cases the flashing rate is roughly once per second with the component values shown. The rate can be decreased by increasing the resistance values, up to a maximum of about 270 k ohms. Both resistors should have equal values. Alternatively, the flashing rates can be increased or decreased by lowering or raising the timing capacitor values. Both capacitors should be given equal values, and can be in the range 1 μF to 1 000 μF.

D.C. lamp chaser projects

Lamp chasers are circuits in which a number of lights are arranged so that they turn on sequentially, with a preset time delay between each operation, until they are all finally on together. *Figure 5.14* shows a practical d.c. chaser of this type.

Normally, with S_1 open, all lamps and s.c.r.s are off, and all capacitors are discharged. When S_1 is first closed power is applied to lamp 1 and to the Q_1 u.j.t. time-delay circuit, and C_1 starts to charge exponentially via R_1. Note at this stage that all s.c.r.s are off, so zero power is applied to the Q_2 or Q_3 networks. After a preset delay C_1 reaches the firing voltage of Q_1, and Q_1 then fires and applies a trigger pulse to the gate of SCR_1, and SCR_1 and lamp 2 turn on.

As lamp 2 turns on it applies power to the Q_2 u.j.t. time-delay network. After another preset delay, therefore, Q_2 fires and turns SCR_2 and lamp 3 on, and lamp 3 applies power to Q_3 and initiates a further timing period which culminates in the firing of SCR_3 and the turning on of lamp 4. The circuit action is then complete, and all lamps are finally on together. The circuit can be extended to incorporate as many lamps as required by simply wiring in a u.j.t. time-delay and an s.c.r. network for each additional lamp stage.

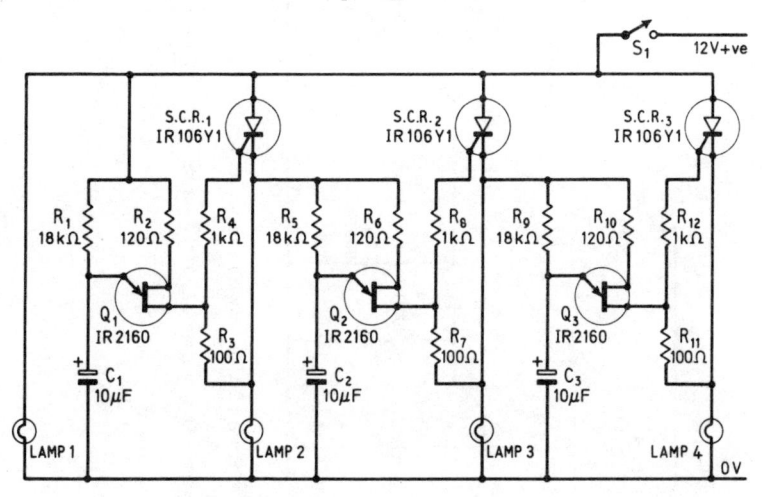

Figure 5.14. D.C. lamp chaser. All lamps are 12 V types with current ratings less than 2 A.

Figure 5.15 shows how the above circuit can be modified to act as a repetitive lamp chaser, which automatically turns off all lamps and then restarts the chaser sequence shortly after the last lamp has turned on. Here, power is connected to the main part of the chaser circuit via normally closed relay contacts $RLA/1$, and the relay is operated from the Q_3 u.j.t. stage via a pulse-expander network and via Q_4 and Q_5. Circuit operation is as follows.

When power is first applied to the circuit via S_1 contacts $RLA/1$ are closed, so power is applied to lamp 1 and to Q_1. After a preset delay Q_1 fires and drives SCR_1 and lamp 2 on, and lamp 2 applies power to Q_2. After another preset delay Q_2 fires and turn on lamp 3 and applies power to Q_3. After a further preset delay Q_3 fires and applies base drive to the Q_4-Q_5 super-alpha amplifier via the $D_3-C_4-R_{12}$ pulse expander network, and relay RLA turns briefly on. As RLA goes on contacts $RLA/1$ open and break the supply to the main part of the circuit, so all lamps turn off and all the timer networks reset. A short time later RLA turns off again through lack of Q_4 base drive, and contacts $RLA/1$ then close again and restart the entire sequence. The process then repeats *ad infinitum*. Diodes D_1 and D_2 are wired in series

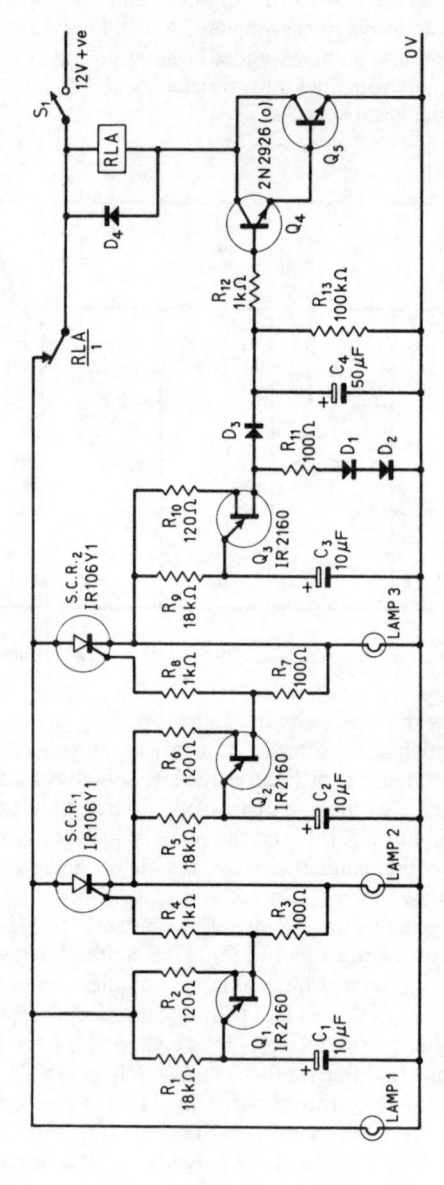

Figure 5.15. Repetitive d.c. lamp chaser. All lamps are 12 V types with current ratings less than 2 A. D_1 to D_4 = general-purpose silicon diodes. RLA = 12 V relay with a coil resistance greater than 120 Ω: Ω one set of N/C contacts.

with R_{11} in the Q_3 u.j.t. network to counter the forward volt drops of D_3 and the Q_5 base-emitter junction and ensure stable triggering of the 12 V relay.

The time delays of each stage of the *Figure 5.14* and *5.15* circuits are determined by the main timing resistor and capacitor time constants of the u.j.t. stages (R_1 and C_1 in the Q_1 stage), and approximate to 1 second with the component values shown. The time constants can be reduced by lowering the timing resistor values, down to a minimum of 6·8 k ohms, or increased by raising the resistance values, up to a maximum of 500 k ohms.

In the *Figure 5.15* circuit the ON time of the relay is influenced by the value of R_{13}, which acts as a shunt discharge path for C_4, and approximates to 1 second with the R_{13} value shown. The ON time can be reduced to roughly one third of a second by reducing R_{13} to 33 k ohms, or can be increased to roughly five seconds by raising the value to infinity, i.e., by removing R_{13} from the circuit.

Lamp dimmer projects

All the triac circuits shown so far are used to give a simple on/off type of power control in which either full power or zero power is applied to the load. Triacs can, however, also be used to give very efficient variable power control of a.c. circuits by using the phase triggering technique described in Chapter 1, and such circuits can be used in lamp dimming applications.

Figure 5.16a shows the basic circuit of a simple phase-triggered lamp dimmer. R_1 and C_1 are wired together as a combined variable potential divider and variable phase shift network. The diac is used as a simple trigger device that fires when the C_1 voltage rises to roughly 35 V (in either polarity) and then partially discharges C_1 into the triac gate, thus triggering the triac on. The diac turns off automatically when the C_1 voltage falls below 30 V or so.

When R_1 is set to a very low value negligible potential divider action or phase shifting takes place, and the C_1 voltage closely follows that of the a.c. power line until the trigger voltage of the diac is reached, at which point the triac fires and turns on the lamp and removes all drive from the $R_1 - C_1$ network. The triac thus fires shortly after the start of each half-cycle under this condition, and almost full power is applied to the load.

When R_1 is set to a very high value, on the other hand, the potential divider action is such that the peak voltage on C_1 only just reaches the 35 V needed to trigger the diac, and the phase shift of C_1 is close to 90°. Since the peak of a half-wave occurs 90° after the start of the half-cycle, the net effect of the low voltage and near-90° phase shift on C_1 is to delay the firing of the triac by about 170°. Under this condition, therefore, the triac does not fire until 10° before the end of each half-cycle, and negligible power is applied to the load. Thus, the R_1-C_1 and diac network enables the firing of the triac to be delayed between roughly 10° and 170° in each half-cycle, and efficient variable power control is available.

The basic circuit of *Figure 5.16a* must be modified slightly before it can be used as a practical lamp dimmer. Since the triac switches from off to on very sharply, and switches fairly high currents, the switching waveform is very rich in harmonics. In the basic circuit these harmonics

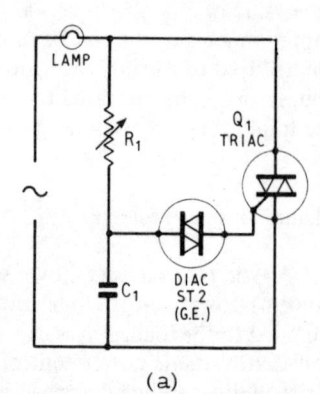

Figure 5.16a. Basic lamp dimmer circuit.

(a)

are fed into the a.c. power line and may cause conducted r.f.i. on a.m. radios. One modification that the circuit needs, therefore, is the incorporation of an r.f. filter to keep the higher harmonics out of the supply line.

A further modification must be made to ensure that, when R_1 is set near its minimum value, the charge currents flowing to C_1 via R_1 are not so large that they damage R_1. This protection can be obtained by wiring a limiting resistor in series with R_1.

Figure 5.16b shows the practical circuit of a simple lamp dimmer incorporating these modifications. R_2 is the charge-current limiting resistor, and L_1-C_2 form the r.f.filter. L_1 must have a current rating greater than that of the lamp, and L_1 can be home-made by scramble winding roughly 100 turns of suitably rated insulated wire on a ½-inch-diameter former. If desired, the body of C_2 can be used as the coil former, providing that C_2 is of a reasonable physical size.

The *Figure 5.16b* circuit makes a useful lamp dimmer, but has one mildly annoying characteristic. R_1 has considerable hysteresis or backlash. If the lamp finally goes fully off when R_1 is increased to 250 k ohms (in the 120 V circuit), it will not start to go on again until R_1

is reduced to about 200 k ohms, and it then burns at a fairly high brightness. The cause of this characteristic is as follows.

Suppose that the triac is fully off and R_1 value is reduced to 201 k ohms, at which point C_1 voltage is not quite reaching the 35 V needed to fire the diac and triac. Under this condition C_1 is discharging by

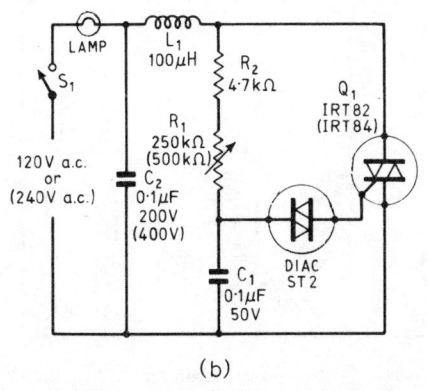

(b)

Figure 5.16b. Practical circuit of a simple lamp dimmer.

nearly 35 V and charging by nearly 35 V on each half-cycle, giving a voltage change of nearly 70 V per half-cycle.

Suppose now that R_1 is reduced to 200 k ohms, and C_1 potential just reaches 35 V and causes the triac to trigger via the diac about 10° before the end of the first operative half-cycle, thus applying very little power to the lamp. As the diac fires in this first half-cycle it reduces the C_1 charge by 5 V, to only 30 V. Consequently, on the following half-cycle C_1 has to discharge 30 V and charge 35 V to trigger the triac, giving a total voltage change of 65 V in the half-cycle. Since R_1 is still set at 200 k ohms, therefore, C_1 reaches 35 V earlier in the second half-cycle than it did in the first, and the triac thus fires well before the end of the half-cycle and applies considerable power to the lamp, causing it to burn at a fairly high level. On all subsequent half-cycles the circuit operates under the same condition as in the second half-cycle, and the lamp continues to burn at a fairly high level.

Once the circuit is operating and C_1 is going through a 65 V change in each half-cycle the R_1 value can be increased to about 249 k ohms before the lamp turns fully off, and the lamp brightness can be reduced to very low levels. The backlash of R_1 and the high initial turn-on brightness level of the lamp are thus caused by the relatively large instantaneous voltage changes that occur on C_1 as the diac fires in the first operating half-cycle.

The backlash and associated problem can be reduced or eliminated in a number of ways. *Figure 5.17* shows an improved lamp dimmer that uses R_3 to reduce the magnitude of the voltage change on C_1 as the diac fires, thus giving a reduction in the backlash effect.

Figure 5.18 shows a circuit that gives a further reduction in backlash. The design is similar to that of *Figure 5.16b*, except that the charge of C_1 is fed to slave capacitor C_3 via the relatively high resistance of R_3.

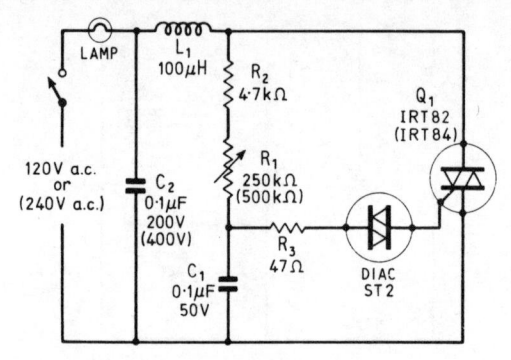

Figure 5.17. Improved lamp dimmer circuit.

C_1 is at a slightly higher voltage than C_3, and C_3 fires the diac once its voltage reaches 35 V. Once the diac has fired it reduces the C_3 potential briefly to 30 V, but C_3 then partially recharges via C_1 and R_3. Little net change takes place in the C_3 voltage as a result of the diac firing, and the circuit thus gives very little backlash. The backlash can be reduced

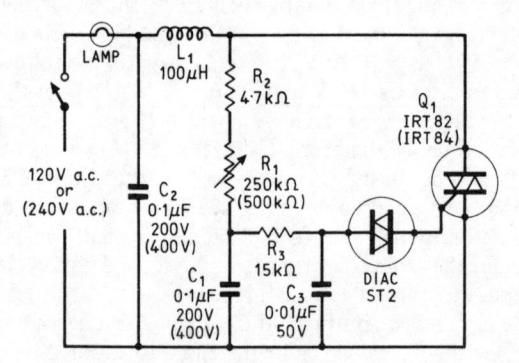

Figure 5.18. Improved lamp dimmer with gate slaving.

to almost zero, if required, by wiring a further resistor in series with the diac to limit the C_3 discharge voltage, as shown in *Figure 5.19*.

The individual diac and triac of the *Figure 5.16* to *5.19* circuits can be replaced by a single quadrac, if required, as shown in *Figure 5.20* to *5.22*. The quadracs used in these three circuits have factory attached

25 LAMP-CONTROL PROJECTS 89

heat radiators, and can handle r.m.s. currents up to 2·2 A at ambient temperatures of 25°C. They can thus handle maximum load powers of 264 W on 120 V lines and 528 W on 240 V lines. These quadracs can, however, handle r.m.s. currents up to 6 A if their case temperatures are kept below 75°C with the help of extra heat sinking, thus allowing maximum load powers of 720 and 1 440 W.

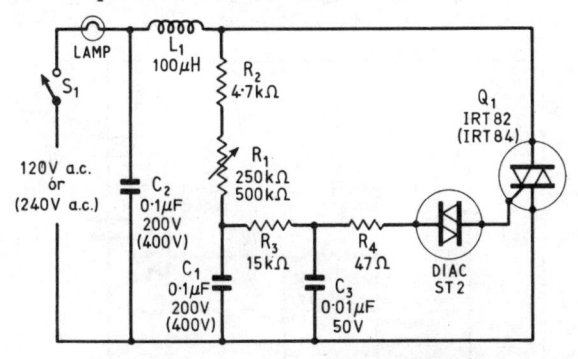

Figure 5.19. Minimum-backlash lamp dimmer.

Another type of lamp dimmer is shown in *Figure 5.23*. This circuit is that of a high-performance zero-backlash lamp dimmer that uses a unijunction transistor as the phase-control element. The u.j.t. (Q_5) applies time-delayed d.c. pulse triggering to the triac gate, and is itself gated on synchronously via Q_4 and line-driven zero-crossing detector Q_2 and Q_3. Q_4 and Q_5 are powered from a 12 V d.c. supply derived from the a.c. line via $R_1-D_1-ZD_1$ and C_1.

The circuit action is such that Q_4 is gated on and Q_5 starts into a timing period shortly after the start of each half-cycle, and a short

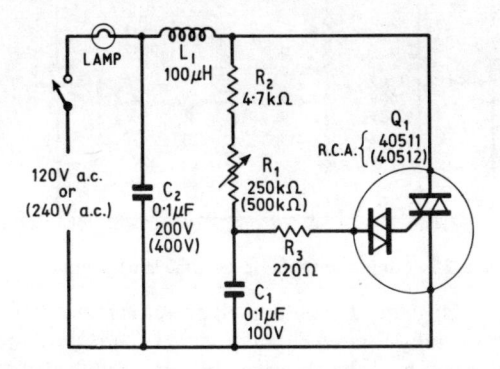

Figure 5.20. Simple quadrac lamp dimmer.

time later the u.j.t. fires and triggers the triac on for the remaining part of the half-cycle. At the end of the half-cycle the triac turns off, and simultaneously the Q_2-Q_3 zero-crossing detector removes base drive from Q_4 and causes the u.j.t. network to discharge C_2 and reset. This process repeats on each half-cycle.

The pulse triggering delay of the circuit can be varied from less than 200 μs to more than 10 ms, and is determined solely by the R_6 setting.

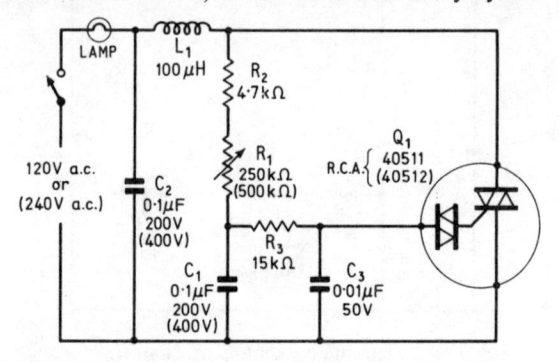

Figure 5.21. Quadrac lamp dimmer with gate slaving.

Since the period of each half-cycle of line voltage is fixed, these time delays can be translated directly into terms of phase delay, and correspond to delays of approximately 3° to more than 180°. The circuit thus gives full lamp dimming control with zero backlash. Note in the

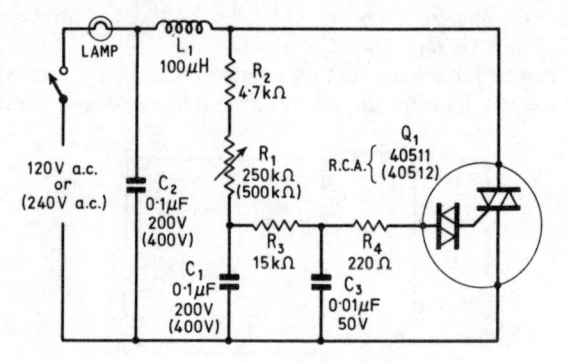

Figure 5.22. Minimum-backlash quadrac lamp dimmer.

circuit that S_1 is ganged to R_6, so that all power is removed from the circuit when R_6 is fully rotated beyond its maximum resistance position.

Figure 5.24 shows how the above circuit can be modified so that it acts as a slow-start lamp-control unit. When S_1 is first closed the lamp intensity builds up fairly slowly from zero to maximum, taking roughly two seconds to reach full brilliance. The circuit is designed to eliminate high turn-on inrush lamp current, and thus to extend lamp life. Circuit operation is as follows.

When power is first applied to the circuit via S_1 capacitor C_4 is fully discharged and acts like a short circuit. Under this condition C_2 charges via high-value resistor R_5 only, so the u.j.t. develops a long time delay and the lamp burns at a very low intensity. As soon as S_1 is closed C_4 starts to charge exponentially via R_6 and D_2, and an exponentially rising voltage appears at the R_6-D_2 junction and partially charges C_2 via D_3 in each operating cycle, thus progressively reducing the u.j.t. timing

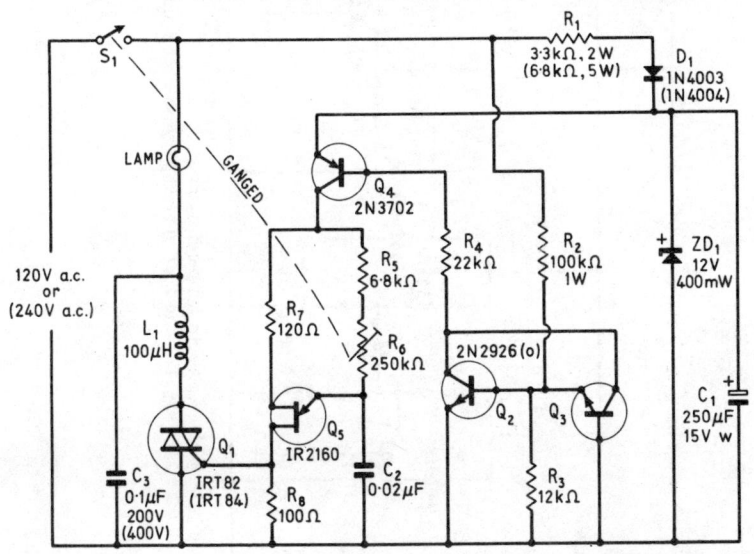

Figure 5.23. High-performance zero-backlash lamp dimmer.

period. Eventually, after a preset delay determined by R_6 and C_4, C_4 becomes fully charged and acts like an open circuit. Under this condition, therefore, C_2 charges in each operating cycle via the parallel combination of R_5 and R_6, and a short delay period is developed and the lamp burns at full brilliance.

Thus, the lamp brilliance of the circuit rises smoothly from zero to maximum when S_1 is first closed, and takes roughly two seconds to reach full intensity. R_5 determines the minimum turn on brightness of the lamp. R_6 determines the maximum brightness of the lamp and, in conjunction with C_4, determines the units starting time constant. Diode D_2 is used to prevent C_4 discharging into the u.j.t. each time Q_5 fires,

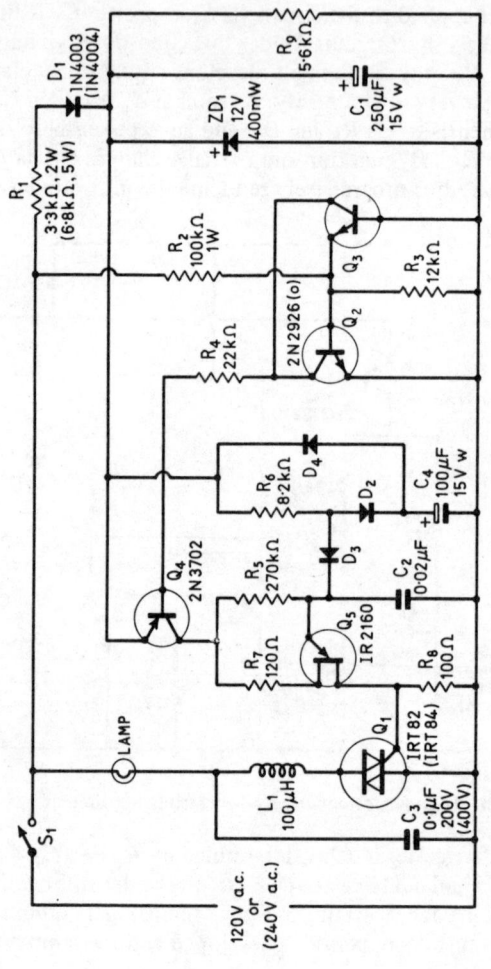

Figure 5.24. Slow-start lamp-control unit. D_2, D_3, and D_4 = general-purpose silicon diodes.

25 LAMP-CONTROL PROJECTS   93

and D_4 automatically discharges C_4 via R_9 and resets the network when power is removed from the circuit via S_1.

Finally, *Figure 5.25* shows how the circuit can be further modified so that it acts as a combined lamp dimmer and slow-start circuit. The

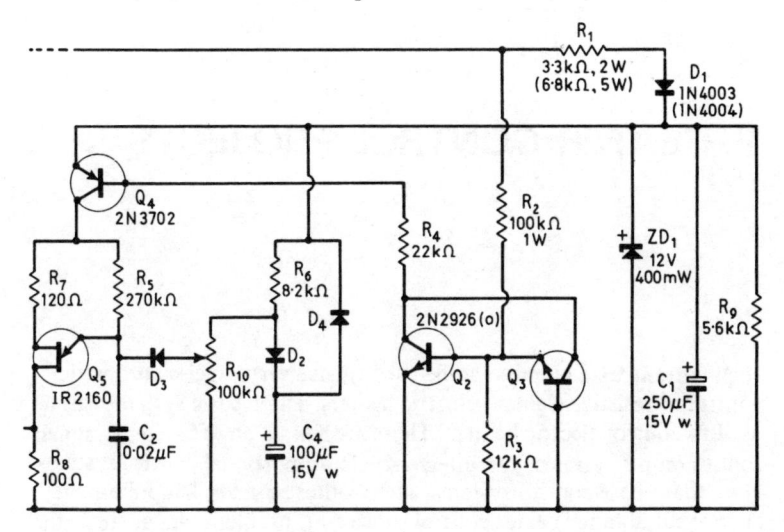

Figure 5.25. Combined lamp dimmer and slow-start circuit.
D_2, D_3, and D_4 = general-purpose silicon diodes.

circuit operates in a similar way to that of *Figure 5.24*, except that the exponential C_4 voltage is fed to C_2 via potential divider R_{10} and diode D_3, thus enabling the maximum lamp intensity to be varied from zero to maximum while at the same time retaining the slow-start characteristic when S_1 is first closed.

CHAPTER 6

15 HEATER-CONTROL PROJECTS

High-current triac circuits can be used to give very precise automatic control of resistive-element electric heaters. Three basic systems can be used to control electric heaters. These are simple on-off systems, synchronous on-off systems, and integral-cycle proportional control systems. Note that phase-control systems, such as those used in lamp dimmers, are not suitable for heater control, since r.f.i. problems are acute at the high power levels involved.

Fifteen heater-control projects of different types are described in this chapter. These projects are designed around the 40575 15 A 200 V triac manufactured by R.C.A., or the IRT84 10 A 400 V triac manufactured by International Rectifier. The 40575 can control heater powers up to 1 800 W on 120 V lines, and the IRT84 can control 2 400 W on 240 V lines.

Simple on-off control projects

Figures 6.1a and *6.1b* show the circuits of two simple heater-controllers with thermostat regulation. At low temperatures the thermostat is closed, so the triacs are gated on via R_1. When the desired room temperature is reached the thermostat opens and removes the gate drive, so the triacs and heaters automatically turn off. The circuits can be turned on and off manually, if desired, via S_1.

The Figure 6.1a circuit uses line-derived gate drive, and thus generates continuous low-level r.f.i. when *TH* is closed. The *Figure 6.1b* circuit uses d.c. gate drive, and does not generate r.f.i. while *TH* is closed. Both circuits may, however, generate a single large pulse of r.f.i. at the instant that the thermostat first closes, as described in Chapter 1.

15 HEATER-CONTROL PROJECTS 95

Figure 6.2 shows the circuit of a simple thermistor-controlled heater. Gate drive is derived from the low-voltage d.c. supply via R_5 and Q_2. This supply also provides power to the temperature-sensitive Q_3 network which controls Q_2. TH_1 is a negative temperature coefficient thermistor, and is wired in a bridge network together with $R_1 - R_2 - R_3$

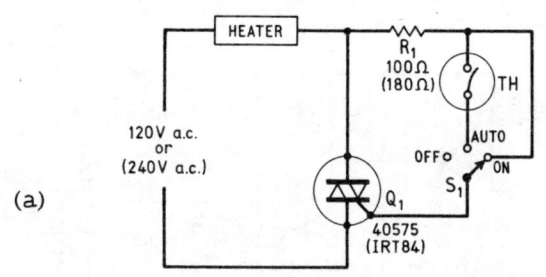

(a)

Figure 6.1a. Simple thermostat-controlled heater.

and R_4. R_1 is adjusted so that the bridge is close to balance at the desired turn-off temperature, with the base-emitter junction of Q_3 slightly forward biased under this condition. Q_3 is used as a bridge-balance detector, and provides drive to Q_2.

Circuit operation is fairly simple. At low temperatures the bridge goes out of balance in such a way that Q_3 and Q_2 are driven heavily on,

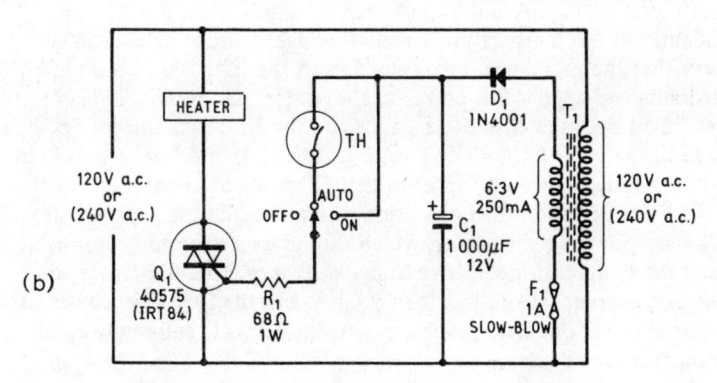

(b)

Figure 6.1b. Improved thermostat-controlled heater.

so high gate drive is applied to the triac and the heater is on. At high temperatures the bridge goes out of balance in such a way that Q_3 and Q_2 are both cut off, so zero gate drive is applied to the triac, and the heater is off. Finally, when temperatures are close to the preset level Q_3 and Q_2 are partially driven on, and the magnitude of the gate drive to the triac is proportional to the difference between the actual and the preset turn-off temperature. The operating mode of the triac and the heater under this condition depends on the magnitude of the gate drive current, as follows.

It will be remembered from Chapter 1 that a triac has two distinct levels of gate trigger sensitivity. The triac in the *Figure 6.2* circuit is

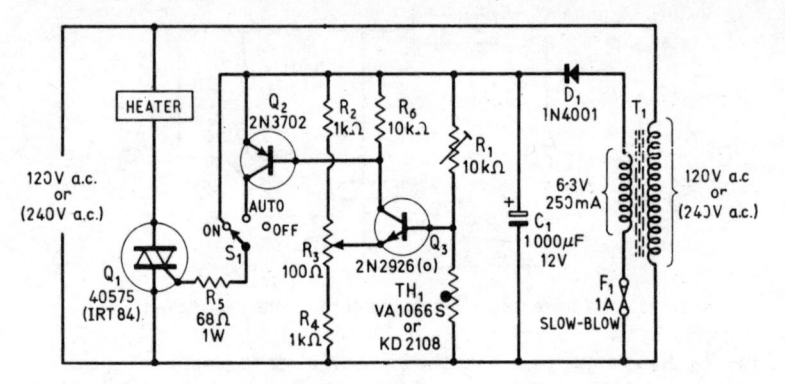

Figure 6.2. Simple thermistor-controlled heater with 'Tri-mode' operation.

gated alternately in the I^+ and III^+ modes, since the gate current is always positive and the main terminal current is alternating, and has typical sensitivities of 15 mA and 35 mA respectively in these modes. Consequently, if the temperature-sensitive bridge is out of balance in such a way that the gate current exceeds 35 mA the triac is driven on in both quadrants and applies full power to the heater. As the TH_1 temperature rises the bridge goes closer to balance and the triac gate current decreases. When the gate current falls to a value less than 35 mA but greater than 15 mA the triac ceases to trigger in the III^+ mode but continues to trigger in the I^+ mode. Under this condition, therefore, the triac applies only half-wave power to the heater, which thus gives a reduced heat output. When the temperature sensitive bridge goes very close to balance the triac gate current falls to less than 15 mA, and the triac then ceases to trigger in either quadrant, so the triac turns fully off and removes all power from the heater.

Thus, the *Figure 6.2* circuit gives fully automatic on half-wave off 'tri-mode' operation of the heater, and enables room temperatures to be controlled to a high level of accuracy. Since d.c. gate drive is used, the circuit does not generate significant r.f.i. when the heater is on, although it may generate a heavy pulse of r.f.i. at the instant that power is first applied to the circuit.

The thermistor used in this and all other thermistor-regulated circuits shown in this chapter can be any n.t.c. type with a resistance in the range

15 HEATER-CONTROL PROJECTS 97

2 000 to 10 000 ohms at the required operating temperature. The Mullard VA1066S and RCA KD2108 types are suitable.

The *Figure 6.2* circuit is set up by moving R_3 to mid-value and adjusting R_1 so that the heater goes into half-wave operation at the required TH_1 temperature. R_3 then acts as a variable-temperature control. TH_1 should be placed remote from the main unit and positioned so that it responds to the mean room temperature.

Synchronous on-off control projects

The simple control circuits of *Figure 6.1* and *6.2* give excellent automatic control of electric heaters, but are prone to generating a heavy pulse of r.f.i. when they are first switched on. In addition, the *Figure 6.1a* circuit generates continuous low-level radiated r.f.i. when the heater is on. Both types of r.f.i. can be eliminated by using the synchronous 'zero-voltage' switching technique, in which gate drive is applied to the triac only in the brief periods when the instantaneous line voltage is close to the zero-voltage cross-over points in each half-cycle. *Figure 6.3* shows the practical circuit of a synchronous thermostat-controlled heater regulator of this type.

A full description of the operating principles and circuit details of this unit are given in Chapter 2. Briefly, however, Q_2 and Q_3 are wired as a zero-crossing detector that is driven from the a.c. power line, and the outputs of Q_2 and Q_3 provide a control signal to the Q_4–Q_5 triac-

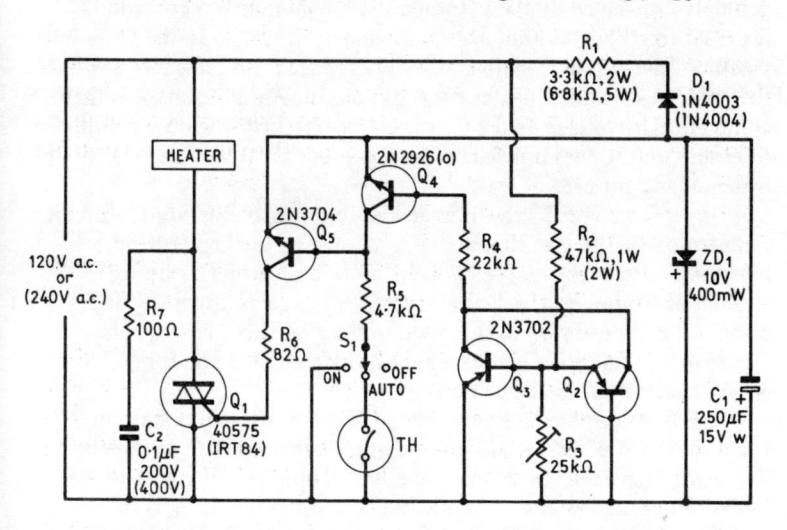

Figure 6.3. Synchronous thermostat-controlled heater.

gate-drive network. Q_4 and Q_5 are powered from a 10 V d.c. supply derived from the power line via $R_1 - D_1 - C_1$ and zener diode ZD_1.

The circuit action is such that a heavy pulse of gate drive is applied to the triac only when the instantaneous line voltage is close to zero at the cross-over points near the beginning and end of each half-cycle, and when base drive is available to Q_4 via R_5. The circuit thus gives r.f.i.-free operation of the heater, and can be inhibited or turned off by simply removing the R_5 current. In *Figure 6.3* thermostat *TH* is wired in series with R_5, and thus gives automatic on-off control of the heater.

At low temperatures the thermostat is closed and the circuit gives synchronous operation of the heater. At high temperatures the thermostat is open and the circuit is inhibited, so the heater is off. The circuit can be turned on and off manually, if required, via S_1.

The *Figure 6.3* circuit is very easy to set up, since R_3 is the only adjustable component in the design (apart from the thermostat). It should be noted, however, that (as pointed out in Chapter 2) this component must be adjusted to suit the particular resistive load that is used with the triac. If a multi-value load, such as a two or three-bar heater, is used, the R_3 adjustment must be made with the load in its minimum-load position, i.e., with only one of the heater bars turned on. Once the initial R_3 adjustment has been made the circuit will function correctly in all positions of the multi-value load.

The R_3 adjustment is in practice very simple. The selected heater is simply connected in place (turned to its minimum load position), R_3 is set to give maximum resistance, and S_1 is turned to the ON position. The R_3 value is then gradually reduced until it is just past the point at which the heater turns full on; the R_3 adjustment is then complete. The voltage across C_1 should be checked when R_3 is adjusted, to ensure that it does not fall below the nominal 10 V value. The circuit is then ready for use.

Figures 6.4 and *6.5* show how the basic synchronous control circuit can be modified to give thermistor regulation of the heater. In the *Figure 6.4* circuit $R_8 - R_9 - R_{10} - R_{11}$ and thermistor TH_1 are wired as a temperature-sensitive bridge, and Q_6 is wired as a simple bridge-balance detector similar to that used in the *Figure 6.2* circuit. The *Figure 6.5* circuit is similar, except that differential amplifier $Q_6 - Q_7$ is used as the bridge-balance detector.

In both cases the circuit action is such that Q_6 is driven hard on when room temperatures are low, in which case current is available to R_5, so the triac turns on synchronously and applies full power to the heater. When temperatures are high, on the other hand, Q_6 is fully cut off and zero current flows in R_5, so the circuit is inhibited and zero power is fed to the heater. Finally, when temperatures are close to the

15 HEATER-CONTROL PROJECTS 99

preset value Q_6 is driven partially on, and the magnitudes of both the R_5 and the triac-gate-drive currents are proportional to the difference between the actual and the preset temperatures. The triac is gated alternately in the I⁻ and III⁻ modes, so tri-mode operation of the heater is given under this condition. The circuits thus give excellent

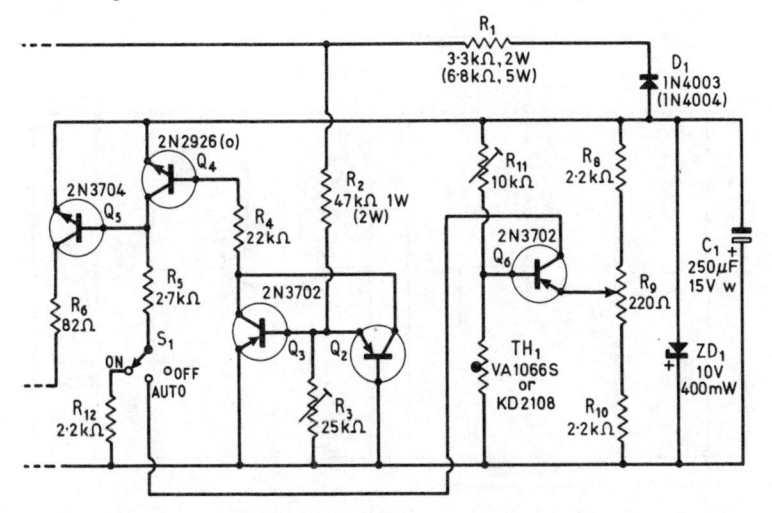

Figure 6.4. Synchronous thermistor-controlled heater.

regulation of room temperature. The regulation of the *Figure 6.5* circuit is slightly better than that of *Figure 6.4*, since the operating point of *Figure 6.4* is influenced by thermally generated shifts in the V_{be} potential of Q_6, while the *Figure 6.5* differential circuit is unaffected by such changes.

The procedure for setting up the two circuits is identical. First, connect the heater in place, set S_1 to the ON position, and then adjust R_3 in the same way as described for the *Figure 6.3* circuit. Next, set R_9 to mid-value, raise TH_1 to the required turn-off temperature, and then adjust R_{11} so that the heater goes into half-wave operation under this condition. All adjustments are then complete, and the circuits are ready for use. R_9 enables the turn-off temperature to be varied a few degrees about the value preset by R_{11}.

Finally, *Figure 6.6* and *6.7* show how the PA424 integrated circuit described in Chapter 2 can be used for synchronous on-off heater-control. In the *Figure 6.6* circuit a thermostat is coupled to the i.c.s differential amplifier input via $R_2 - R_3$ and R_4, and the circuit action is such that the triac is turned on synchronously when the thermostat is closed, but is off when the thermostat is open.

15 HEATER-CONTROL PROJECTS

In the *Figure 6.7* circuit a thermistor and variable resistor are wired to the amplifier input, and the circuit works in the same way as that of *Figure 6.5*. The triac is turned on synchronously at low temperatures, is off at high temperatures, and is operated in the synchronous tri-mode at temperatures very close to the pre-set value.

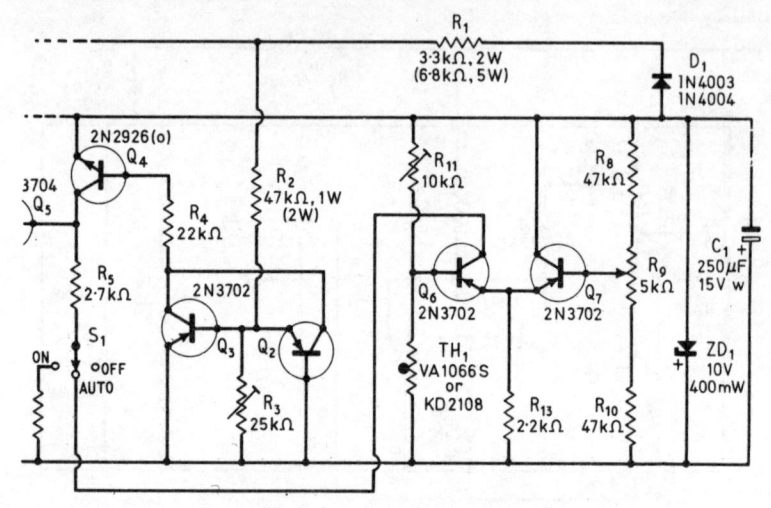

Figure 6.5. Alternative synchronous thermistor-controlled heater.

Integral-cycle variable-power projects

All the circuits that we have looked at so far give a simple on-off control of electric heaters, and thus give only a coarse control of heat output. Fully variable r.f.i.-free control of heater output can be

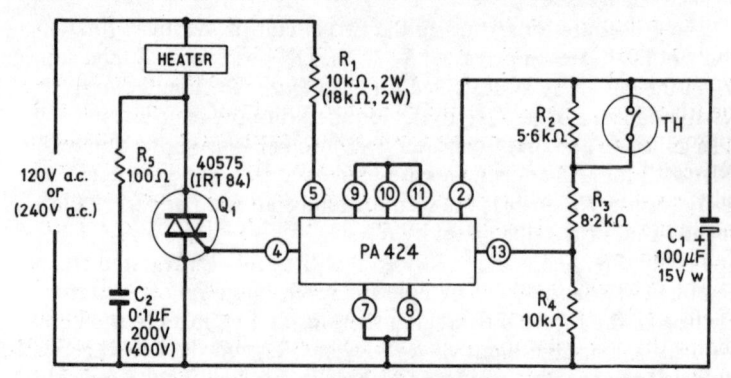

Figure 6.6. Synchronous thermostat-controlled heater using i.c.

obtained by using a synchronous integral-cycle triac switching technique, in which power is applied to the heater for only a definite integral number of half-cycles out of (say) every one hundred. Thus, if power is applied for only fifty half-cycles in each hundred the heater will operate at 50 per cent of full power, and if power is applied for ninety half-cycles in every hundred it will operate at 90 per cent of full power, and so on.

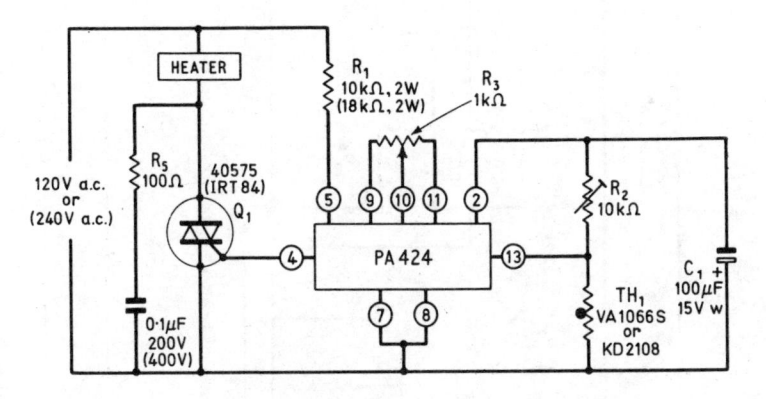

Figure 6.7. Synchronous thermistor-controlled heater using i.c.

Figure 6.8 shows the practical circuit of an integral-cycle variable-power heater controller. Circuit operation is fairly simple. $Q_2 - Q_3 - Q_4 - Q_5$ are wired as a normal synchronous controller, and $Q_6 - Q_7$ are wired as an astable (free-running) multivibrator that applies a rectangular on-off control signal to R_5. The astable action is such that Q_6 and Q_7 switch on and off repetitively in opposition to one another. When Q_6 is on, base drive is applied to Q_5 via R_5, and the triac is turned on synchronously. When Q_6 is off all base drive is removed from Q_5, and the triac is off. The astable multivibrator has a total cycling period of approximately one second, and its mark-space ratio is fully variable between 11 : 1 and 1 : 11 via R_{12}. Consequently, the heater power is fully variable between approximately 8 per cent and 92 per cent of maximum via R_{12}. The heater can be turned fully off or fully on via S_1, if required.

The manually operated variable-power facility can be readily added to any of the synchronous control circuits described earlier by simply adding the multivibrator and a suitable selector switch to these designs, so that they can be made to give either automatic or manual control via the selector switch. *Figures 6.9* and *6.10* show how the circuits of

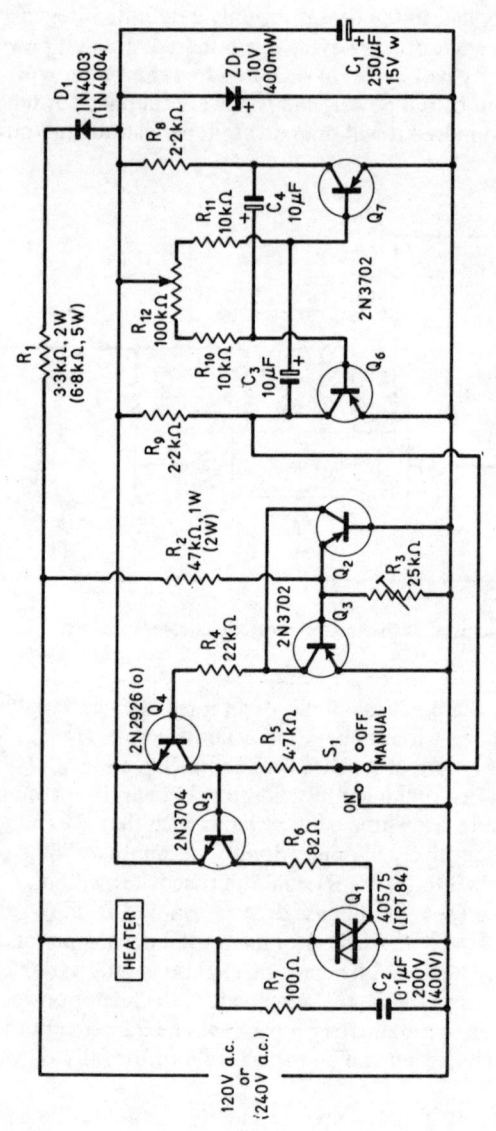

Figure 6.8. Integral-cycle heater controller, manually controlled.

15 HEATER-CONTROL PROJECTS   103

Figures 6.3 and *6.4* can be suitably modified. *Figure 6.11* shows an i.c. version of the manually operated heater controller, and *Figures 6.12* and *6.13* show how the facility can be added to the *Figure 6.6* and *6.7* circuits.

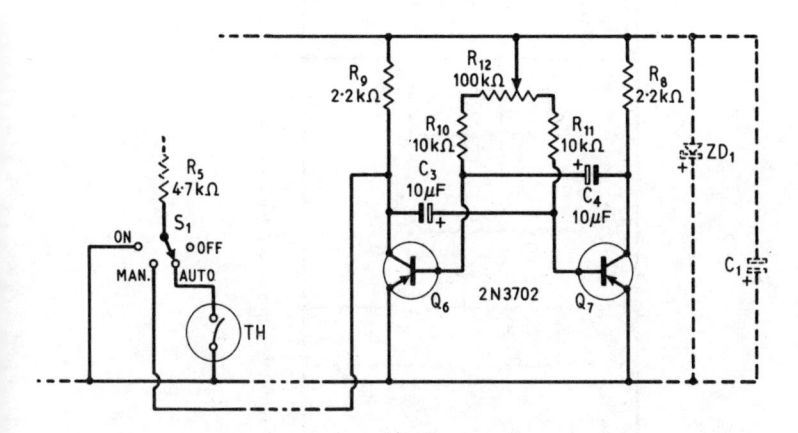

Figure 6.9. Modifications for adding variable-power facility to *Figure 6.3* circuit.

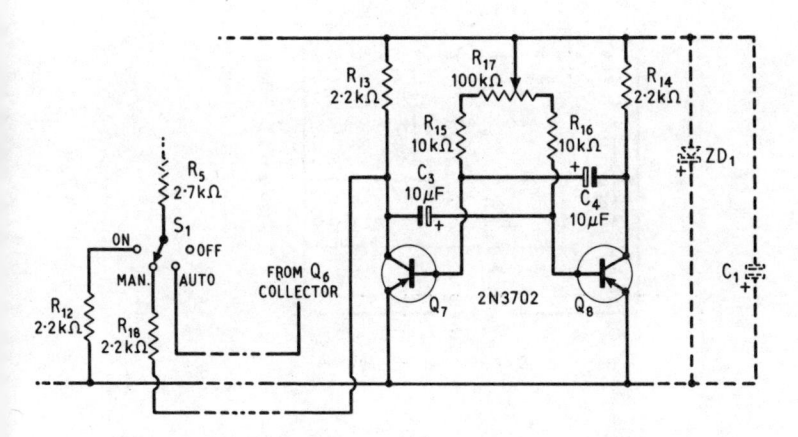

Figure 6.10. Method of adding variable-power facility to *Figure 6.4* circuit.

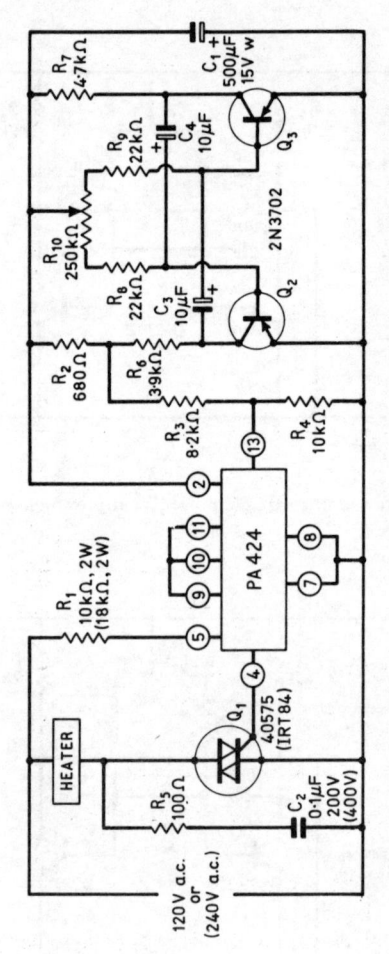

Figure 6.11. Manually controlled integral-cycle heater controller using i.c.

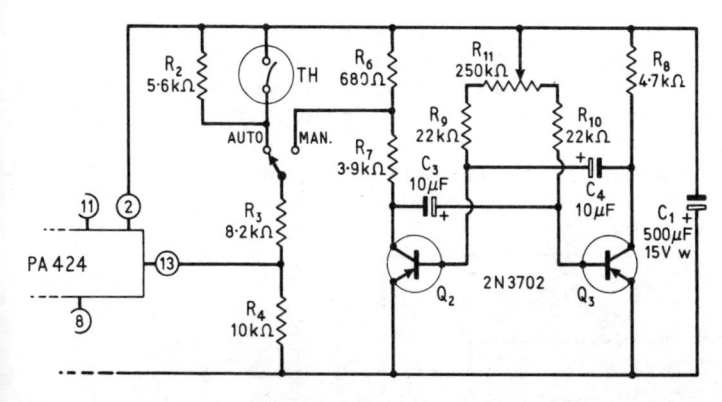

Figure 6.12. Method of adding variable-power facility to *Figure 6.6* circuit.

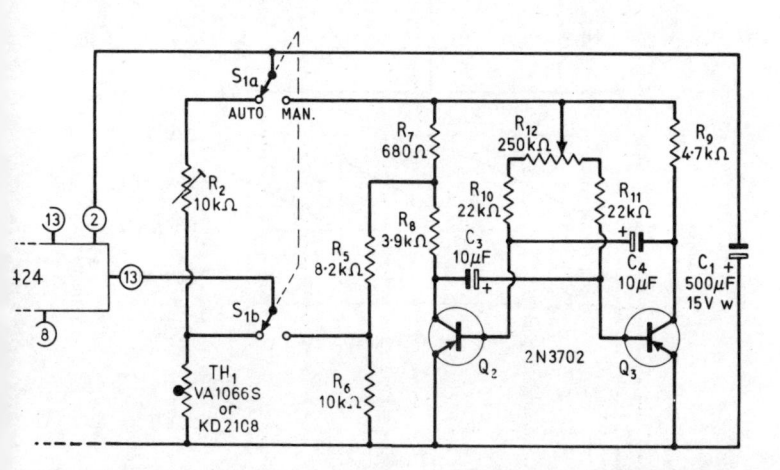

Figure 6.13. Method of adding variable-power facility to *Figure 6.7* circuit.

Automatic integral-cycle control projects

Thermistor-regulated synchronous circuits can be designed to give automatic integral-cycle control of electric heaters. Such circuits give very accurate regulation of room temperatures. The operating principle of a self-regulating integral-cycle heater controller can be understood with the aid of *Figure 6.14*. *Figure 6.14a* shows the basic circuit of the temperature-sensing part of the system. $R_1-R_2-R_3$ and TH_1 are connected as a temperature-sensitive bridge, and Q_1-Q_2 are wired as a differential balance detector that can apply an inhibit signal to a synchronous on-off heater controller via Q_1 collector. A repetitive

106   15 HEATER-CONTROL PROJECTS

sawtooth waveform, with an amplitude of 300 mV and a period of one second, is applied to Q_2 base via C_1, and the circuit action is such that an inhibit signal is fed to the synchronous on-off circuit whenever Q_1 turns off as the instantaneous voltage at point 'B' goes negative to that at 'A'.

Figure 6.14b shows the voltages that appear at points 'A' and 'B' under different temperature conditions when the circuit is set to

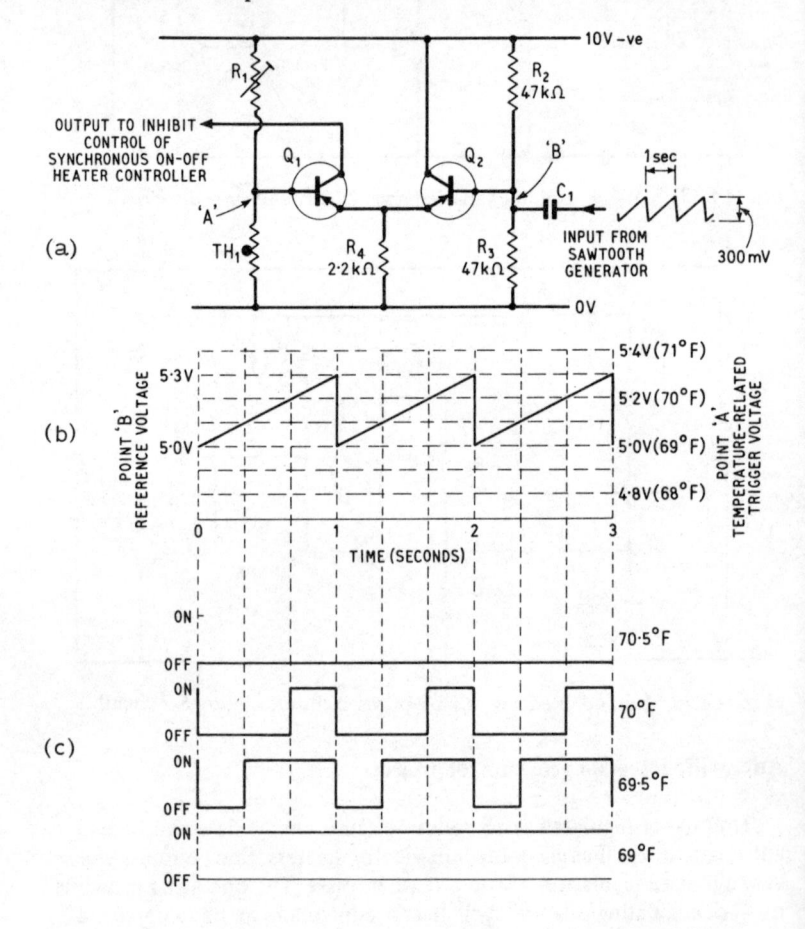

Figure 6.14a. Temperature-sensitive section of automatic integral-cycle heater controller.

Figure 6.14b. Voltages at points 'A' and 'B' of the *Figure 6.14a* circuit.

Figure 6.14c. Heater-power waveforms developed by the automatic integral-cycle controller at various temperatures.

maintain a room temperature of 70°F, and *Figure 6.14c* shows the resulting heater output levels at four different temperatures. It can be seen that a low amplitude sawtooth waveform is superimposed on a fixed reference potential of 5 V at point 'B' of the circuit, and that a steady potential appears at point 'A' but has an amplitude that varies with temperature. R_1 is adjusted so that its resistance is slightly greater than that of TH_1 at 70°F, so that a potential of 5·2 V appears at point 'A' under this condition.

Thus, when the room temperature is below 69°F the TH_1 resistance is high and point 'A' is always negative to point 'B', so Q_1 is permanently biased on and full power is applied to the heater, as shown in *Figure 6.14c*. As the room temperature rises the TH_1 resistance decreases, and the potential at point 'A' falls. Consequently, the circuit passes through an area (between 69°F and 70·5°F) where Q_1 is turned on and off once every second by the superimposed sawtooth waveform of point 'B'. When the temperature rises to 69·5°F, Q_1 and the heater is turned off for one third of each one second sawtooth period, so the heater output falls to two-thirds of maximum. At 70°F Q_1 and the heater are turned off for two-thirds of each one second period, so the heater output falls to one-third of maximum. Eventually, when the room temperature rises to 70·5°F, the voltage at point 'A' becomes permanently positive to that at point 'B', so Q_1 and the heater are permanently turned off.

The important point to note about the self-regulating integral-cycle heater control system is that it applies full power to the heater until the room temperature rises to within a degree or so of the preset level, and that the heater output then reduces progressively as the preset level is approached, the heat output being proportional to the thermal requirements of the room. Eventually, when the preset temperature is reached, the heater does not turn fully off, but gives just sufficient output to exactly counterbalance the natural heat losses of the room. The heater only turns fully off when the room temperature is raised slightly above the preset level by an external cause, such as a rise in outside temperature. The system gives very precise regulation of room temperature.

Figure 6.15 shows the practical circuit of a self-regulating integral cycle heater controller. The circuit is similar to that of the *Figure 6.5* synchronous thermistor-controlled heater unit, but has a low-level sawtooth waveform imposed on Q_7 base from unijunction oscillator Q_8. The imposed sawtooth thus causes the circuit to operate in the self-regulating integral-cycle mode. The procedure for initially setting up this circuit is quite simple, and is as follows.

First, connect the selected heater in place, turn S_1 to the ON position, and adjust R_3 in the same way as described for the *Figure 6.5*

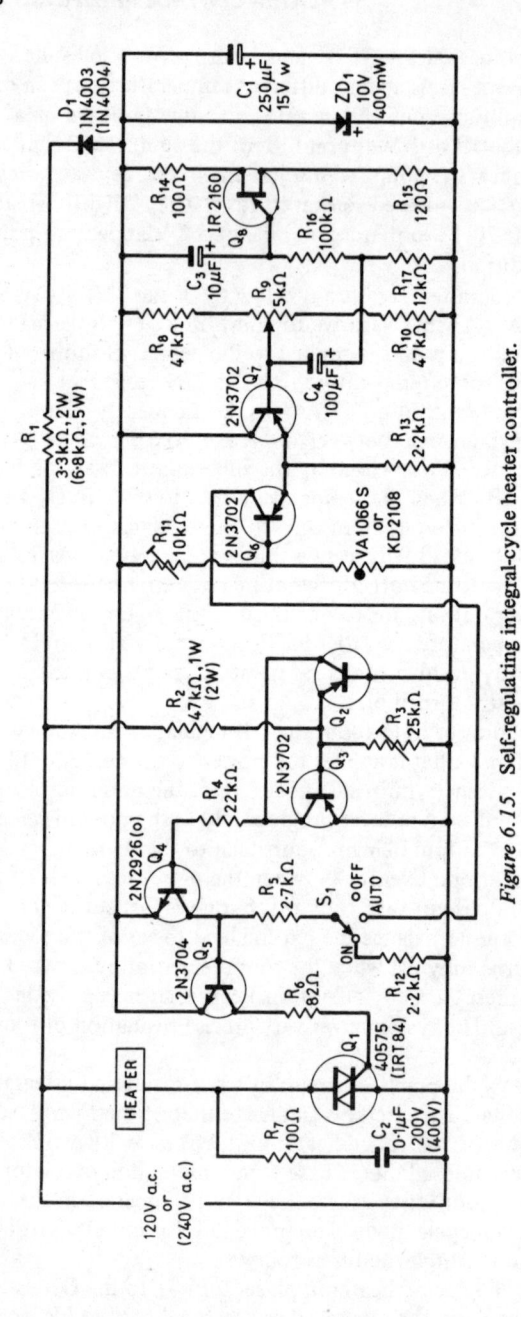

Figure 6.15. Self-regulating integral-cycle heater controller.

circuit. Next, turn S_1 to the AUTO position, set R_9 to mid-value, raise the thermistor to the required 'normal' room temperature level, and then adjust R_{11} so that the heater output drops to roughly one third of maximum. All adjustments are then complete, and the circuit is ready for use. Room temperatures can be varied several degrees about the preset level via R_9.

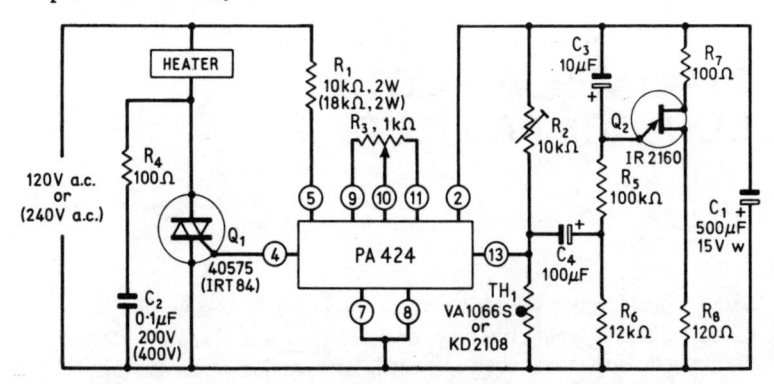

Figure 6.16. Self-regulating integral-cycle heater controller using i.c.

Finally, *Figure 6.16* shows the i.c. version of the basic integral-cycle heater controller. The circuit is developed by simply adding a unijunction sawtooth generator to the circuit of *Figure 6.7*.

CHAPTER 7

15 UNIVERSAL-MOTOR CONTROL PROJECTS

Series-wound 'universal' electric motors (so called because they can operate directly from either a.c. or d.c. power sources) are widely used in domestic appliances such as electric drills and sanders, sewing machines, food mixers, etc. Domestic universal motors are almost invariably operated from a.c. power lines, and are usually designed as single-speed devices.

This speed limitation can be overcome by using phase-controlled triggering techniques in conjunction with an s.c.r., triac, or quadrac. In addition, fairly simple thyristor circuitry can be used to sense variations in motor speed and automatically adjust the power feed to the motor so that the speed remains fairly constant in spite of variations in motor loading. Thyristor control projects for universal-motors are thus of considerable interest, and fifteen such projects are described in this chapter.

Universal-motor characteristics

A conventional universal-motor consists of a simple field winding and an armature wired in series between the motor terminals. When current passes between the motor terminals opposing magnetic fields are set up between the field winding and the armature, and the armature rotates under the resulting magnetic thrust. As the armature rotates in the magnetic field it generates a voltage opposite in polarity to that impressed on the motor terminals. The magnitude of this back-e.m.f. is proportional to the armature (motor) speed. The magnitude of the motor current is proportional to the difference between the applied voltage and this speed-dependent back-e.m.f., and the motor torque is proportional to the motor current.

Study of the above paragraph shows that a universal motor has inherently self-regulating speed characteristics. When a voltage is first applied to the motor terminals its armature is stationary and the back-e.m.f. is zero, so a heavy starting current flows in the motor and produces a high starting torque that applies maximum acceleration and thrust to the armature. The motor thus accelerates to maximum speed very rapidly. When the motor is operating at maximum speed its back-e.m.f. is high, so its current and torque are relatively low. When a mechanical load is applied to the motor its speed tends to decrease. As the speed falls the armature back-e.m.f. decreases, and the difference between the back-e.m.f. and the applied voltage rises. Consequently, the motor current and torque automatically increase, and tend to accelerate the motor back to its original speed. The motor thus exhibits automatic self-regulating speed-control characteristics.

Universal motors can be operated equally well from a.c. and d.c. supplies. This fact enables the motor to be operated from half-wave rectified a.c. supplies, and it is of particular interest to note that, because of its inherent self-regulating characteristics, a half-wave-operated motor develops only 20 per cent or so less speed and power than its full-wave equivalent. This factor enables very efficient and inexpensive half-wave s.c.r. systems to be used for motor control.

A further point of interest is that in half-wave operation the residual magnetism of the motor causes the rotating armature to produce a back-e.m.f. during the non-conducting half-cycles. This back-e.m.f. is proportional to motor speed, and can be sensed by simple electronic circuitry and used to provide a feedback signal to enhance the self-regulating speed control characteristics of the motor.

Half-wave control projects

Half-wave operation of a.c.-driven universal motors results in only a 20 per cent or so reduction of available motor speed and power compared with that available from full-wave operation. Half-wave control systems are thus very attractive from both the economy and efficiency points of view.

Figure 7.1a shows the basic circuit of a simple half-wave variable-speed controller, working on the phase-triggering principle. R_1 and C_1 form a combined variable phase-shift and potential divider network that enables the s.c.r. triggering to be delayed by up to almost 180° in each positive half-cycle. At the start of each positive half-cycle the s.c.r. is off, and power is applied to the R_1-C_1 phase-delay network via D_1. After a preset phase delay the C_1 voltage rises to the firing potential of the trigger device, which then fires and discharges C_1 into the gate of

112   15 UNIVERSAL-MOTOR CONTROL PROJECTS

the s.c.r., which turns on and applies power to the motor and removes the drive from the $D_1-R_1-C_1$ network.

When R_1 is set to its minimum value the R_1-C_1 network provides negligible phase-shift or attenuation, so the s.c.r. fires shortly after the start of each positive half-cycle, and high power is applied to the motor, which operates at high speed. When R_1 is set to its maximum value the R_1-C_1 network provides a large phase shift and high attentuation, so

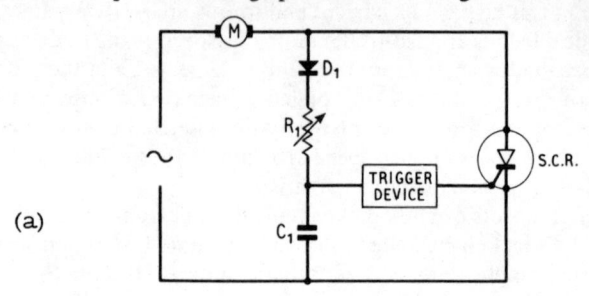

Figure 7.1a. Basic half-wave motor-speed controller.

the s.c.r. triggers on a few degrees before the end of each positive half-cycle, and very little power is applied to the motor, which operates at low speed. Thus, the speed of the motor can be fully varied from maximum to zero by adjusting the R_1 value.

Figure 7.1b shows the practical circuit of a half-wave speed controller, using a s.u.s. as the trigger device. R_2, wired in parallel with R_1, enables the circuit to be set so that the motor speed just falls to near-zero at the maximum resistance setting of R_1, thus giving a maximum R_1 control

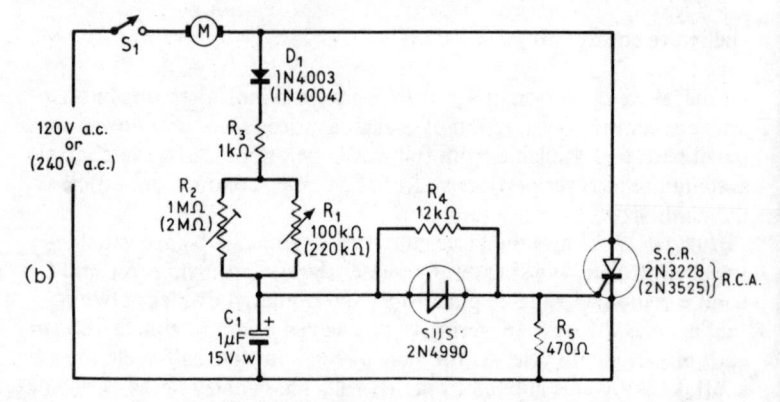

Figure 7.1b. Practical half-wave motor-speed controller, with s.u.s. trigger, for use with motors with current ratings up to 3 A.

15 UNIVERSAL-MOTOR CONTROL PROJECTS 113

range. R_3 is used to limit the peak charging current of C_1 to a safe value when R_1 is set close to zero resistance. R_4 and R_5 provide a discharge path for C_1 on negative half-cycles, and ensure that the circuit operates with negligible backlash.

Figure 7.2 shows how the s.u.s. of *Figure 7.1b* can be replaced by an equivalent transistor trigger network. Q_1 and Q_2 act as a high impedance until the voltage across them rises to roughly 7·5 V, at which point the two transistors regenerate and go into a low impedance state, thus discharging C_1 into the s.c.r. gate. The trigger voltage is determined by the $R_5 - R_6$ ratio.

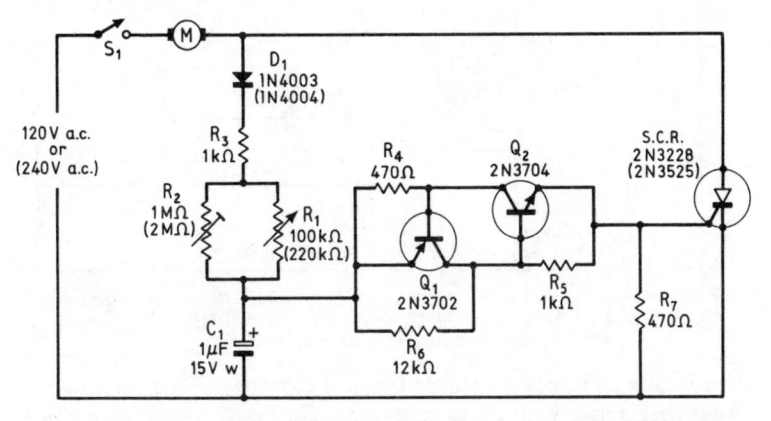

Figure 7.2. Half-wave motor-speed controller, with transistor trigger, for use with motors with current ratings up to 3 A.

Figure 7.3 shows how a degree of self-regulation of speed can be added to the *Figure 7.2* circuit. Here, resistor R_9 is wired in series with the s.c.r. cathode, and a voltage is developed across R_9 when the s.c.r. is on. The peak amplitude of this voltage is directly proportional to the peak amplitude of the motor current. C_2 is charged to the peak value of this voltage via D_2, and the C_2 voltage is imposed on Q_2 base via R_8 and reduces the trigger voltage of the $Q_1 - Q_2$ switch.

Suppose, then, that the motor is set to operate at some particular speed via R_1, and the motor load is suddenly increased. As the load is applied the motor speed tends to decrease, and the motor current rises and the peak R_9 voltage increases. The C_2 voltage also rises and thus reduces the firing voltage of the $Q_1 - Q_2$ switch. The transistor switch and the s.c.r. thus fire earlier on succeeding half-cycles of positive voltage, so the power feed to the motor automatically increases and tends to bring the motor speed back to its original value. A degree of self-regulation of speed thus occurs.

114 15 UNIVERSAL-MOTOR CONTROL PROJECTS

It should be noted that the degree of regulation of the above circuit is not particularly great, since the relationship between the motor current and the feedback power-control section of the circuit is not linear. The performance if the circuit is, however, significantly better than that of the unregulated *Figure 7.2* design. For best results the R_9

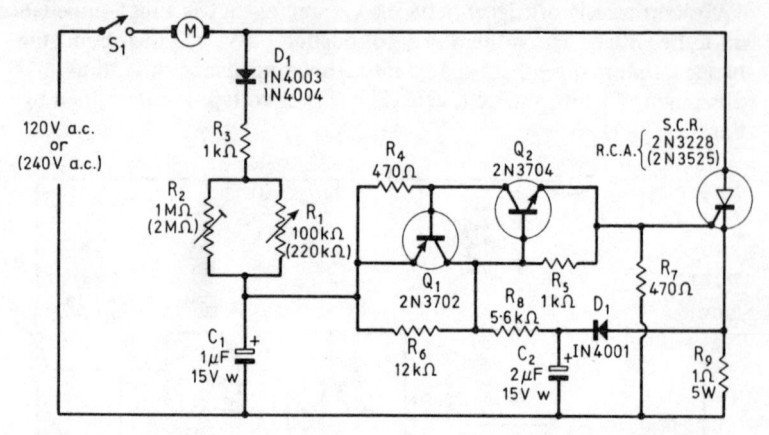

Figure 7.3. Half-wave motor-speed controller, with self-regulation, for use with motors with current ratings up to 3 A.

value may need to be selected by trial and error to suit the particular motor that is used with the circuit.

Figure 7.4 shows the practical circuit of an alternative type of self-regulating speed controller. This circuit senses variations in motor speed via the back-e.m.f. of the armature, and uses these variations to shift the phase-triggering point of the s.c.r. and thus maintains a constant motor speed in spite of variations in motor load. This self-

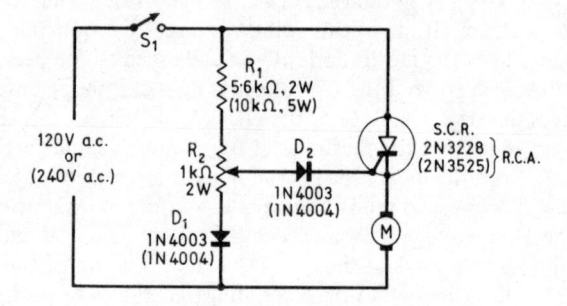

Figure 7.4. High performance self-regulating half-wave speed controller, for use with motors with current ratings up to 3 A.

regulating system is highly efficient, and the circuit is particularly recommended for controlling electric drills and sanders, which are subject to large load variations.

The *Figure 7.4* circuit is deceptively simple. R_1 and R_2 form a simple potential divider, and D_1 acts as a half-wave rectifier. The action of $R_1 - R_2$ and D_1 is such that an attenuated version of each positive half-cycle of line voltage appears on the slider of R_2, and this attenuated voltage is applied to the s.c.r. gate via D_2. At the start of each positive half-cycle the s.c.r. is off, and a speed-dependent voltage is imposed on the s.c.r. cathode by the back-e.m.f. of the motor armature. When the instantaneous line-derived gate voltage exceeds the speed-derived cathode voltage in each positive half-cycle D_2 and the s.c.r. gate become forward biased, and the s.c.r. then triggers on and applies power to the motor for the remaining part of the half-cycle.

Thus, if R_2 is set so that a large fraction of the line voltage appears on R_2 slider the s.c.r. gate becomes forward biased and the s.c.r. turns on and applies power to the motor very early in each positive half-cycle, so high power is applied to the motor, which operates at high speed. When, on the other hand, R_2 is set so that a smaller fraction of the line voltage appears on R_2 slider, the s.c.r. gate becomes forward biased at a later stage in each positive half-cycle, so the s.c.r. applies less power to the motor, which operates at a reduced speed.

It should be noted at this stage that, since the peak voltage of a half-cycle occurs at a phase angle of 90°, the s.c.r. in this circuit can not be triggered via its gate at an angle later than 90° in each positive half-cycle. I.E., when the s.c.r. fires, it always delivers a minimum of one quarter-cycle of power to the motor. This characteristic does not detract from the low-speed performance of the circuit, however, because of a phenomena known as 'skip-cycling'.

Suppose that R_2 is set so that only a small fraction of the line voltage appears on R_2 slider, and that the motor is operating at a low speed under no-load conditions. Since the motor produces a very low back-e.m.f. at a low speed the s.c.r. is triggered on at a low value of R_2-slider voltage, and the s.c.r. then delivers a minimum of a quarter-cycle of power to the motor. This relatively large chunk of power produces high torque and acceleration in the motor, so the motor speed increases rapidly and the back-e.m.f. increases to the point where it exceeds the peak (90°) R_2-slider voltage in the following positive half-cycle.

Consequently, the s.c.r. fails to fire in the next positive half-cycle, and, since the motor is robbed of power drive, the motor speed starts to fall. Eventually, a point is reached where the motor speed and back-e.m.f. fall to a relatively low value again, and the s.c.r. fires once more. The s.c.r. then delivers a minimum of one quarter-cycle of power to the motor, and the whole process repeats. This 'skip-cycling' may occur over one or several half-cycles, depending on the speed setting of R_2. In the latter case the mean motor speed, averaged over a period of a few hundred milliseconds, may be very low, and R_2 thus enables the motor speed to be fully varied from maximum to zero. Since the motor has considerable mass and produces considerable inertia, and since power pulses are delivered to the motor at a rate of tens per second, the full range of motor speeds is provided with very little apparent sign of judder or roughness, and the circuit produces apparently smooth control of speed.

Self-regulation of motor speed occurs because of the circuits ability to sense variations in the speed-dependent back-e.m.f. of the motor, and occurs over the full range of speed control settings. Suppose that R_2 is set so that the s.c.r. is fired late in each positive half-cycle, and the motor is operating at a relatively high speed under no-load conditions, and that a heavy load is suddenly applied to the motor. As the load is applied the motor speed decreases, and the back-e.m.f. of the motor falls. With a reduced back-e.m.f. the s.c.r. gate becomes forward biased earlier in each positive half-cycle, so the s.c.r. triggers on earlier in each half-cycle and delivers increased power to the motor. This increased power raises the motor torque and accelerates the motor speed back towards its original value, and little net change in motor speed occurs.

The self-regulating characteristics of the circuit are particularly good when the motor is operated under low-speed skip-cycling conditions. Suppose that the motor is operating at a low speed and is lightly loaded, and that skip-cycling is occurring over (say) ten half-cycles. The number of skip-cycles are determined by the motors deceleration characteristics, and these characteristics are greatly influenced by motor loading. Thus, if a moderate increase is made in the motor loading the deceleration rate may be doubled, and the motor then skips for only five half-cycles before a further quarter-cycle of power is applied. If the load is further increased the skip-cycling rate reduces further, and if an exceptionally heavy load is placed on the motor the deceleration rate may become so great that skip-cycling ceases and the s.c.r. fires in each positive half-cycle in order to maintain a constant motor speed. The circuit thus exhibits excellent low-speed self-regulating characteristics over a wide range of load variations.

Since the circuit delivers a minimum of a quarter-cycle of power to the motor under skip-cycling conditions the motor produces very high torque at low speed settings. Consequently, the motor produces perceptible judder under conditions of heavy loading at low speeds. Some users find this judder objectionable. The judder can be reduced, at the expense of reduced low-speed torque, by modifying the circuit as shown

15 UNIVERSAL-MOTOR CONTROL PROJECTS   117

in *Figure 7.5*. Here, C_1 produces a phase shift in the s.c.r. gating network and enables the minimum firing angle of the s.c.r. to be reduced to less than one quarter-cycle, thus enabling smaller increments of power to be applied to the motor. The C_1 value should be selected (between 1 µF and 10 µF) to give a reasonable compromise between good high-torque and low-judder characteristics at low speeds, the

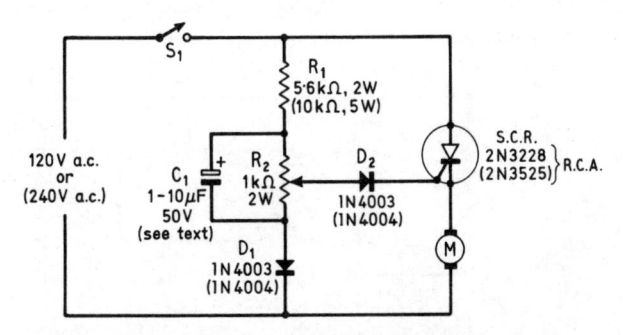

Figure 7.5. High performance self-regulating half-wave speed controller with reduced low-speed judder, for use with motors with current ratings up to 3 A.

compromise being entirely a matter of personal preference. The author's own preference is for the unmodified *Figure 7.4* circuit.

A minor disadvantage of the *Figure 7.4* and *7.5* circuits is that the R_1-R_2 network must supply the full gating current of the s.c.r., and that the s.c.r. must be a fairly sensitive type, otherwise the R_1 and R_2 values must be made so low that they dissipate grossly excessive power (heat). *Figure 7.6* shows a circuit that overcomes this snag and enables a low-cost low-sensitivity s.c.r. to be used as the motor control element.

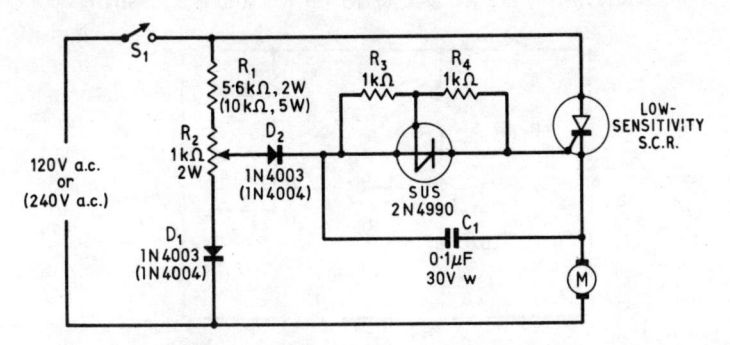

Figure 7.6. High-performance self-regulating half-wave speed controller with s.u.s.-capacitor gate-slaving.

118 15 UNIVERSAL-MOTOR CONTROL PROJECTS

Here, C_1 is used to store a charge voltage proportional to the difference between the R_2-slider and s.c.r. cathode voltage in each half-cycle. When the C_1 voltage exceeds the forward breakover voltage of the s.u.s. (roughly 1·2 V with the component values shown) and that of the s.c.r. gate the s.u.s. regenerates and rapidly discharges C_1 into the s.c.r. gate, thus driving the s.c.r. on. The s.c.r. gate current is thus provided by C_1, which acts as a slaving capacitor, and not directly by R_1 and R_2, which can thus be given sensibly low values.

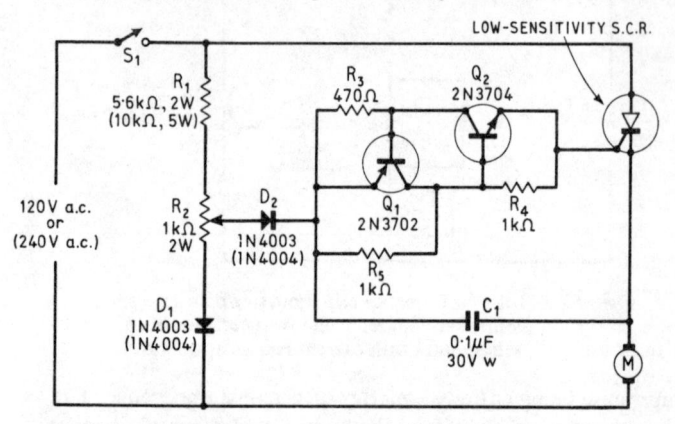

Figure 7.7. High-performance self-regulating half-wave speed controller with transistor-capacitor gate-slaving.

Figure 7.7 shows how the s.u.s. of the *Figure 7.6* circuit can be replaced by a transistor-equivalent regenerative switch. This circuit operates in the same way as that of *Figure 7.6*.

Finally, *Figure 7.8* shows an alternative solution to the sensitivity problem. Here, SCR_1 is used as a 'slave' device, and is a sensitive s.c.r.

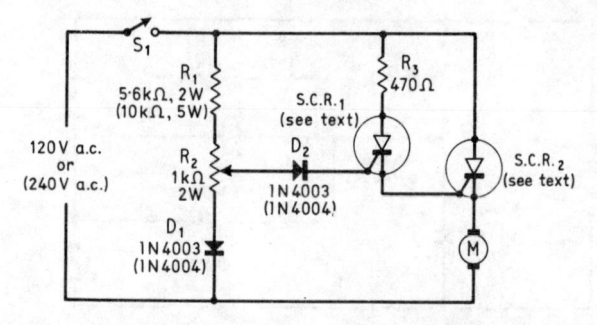

Figure 7.8. High-performance self-regulating half-wave speed controller with s.c.r. gate-slaving.

15 UNIVERSAL-MOTOR CONTROL PROJECTS 119

with a current rating of 1 amp or greater. This s.c.r. is triggered on in the same way as the s.c.r. of *Figure 7.4*, but when it goes on it switches gate drive into the main control s.c.r. via R_3 and the power line, and thus turns on SCR_2, which is a low-sensitivity high-current device. Apart from the differences mentioned, the *Figure 7.6* to *7.8* circuits operate in exactly the same way as that of *Figure 7.4*.

Full-wave control projects

As has already been pointed out, full-wave operation of a universal motor results in only a 20 per cent or so increase in power and speed over that available from half-wave operation. Full-wave speed control circuits thus offer little performance advantage over the simple half-wave designs already described. In addition, full-wave designs are difficult to provide with good self-regulation characteristics. No full-wave design has, in fact, yet been produced that can compete in cost and performance with the simple self-regulating half-wave circuit of *Figure 7.4*.

Full-wave designs are never-the-less widely used, and are definitely useful in unregulated speed-control applications, particularly when a triac or quadrac is used as the motor-control element. Eight practical triac and quadrac designs are presented in this final section of this chapter. All of these designs use the phase-triggering principle, and are simple developments of the lamp-dimmer circuits already described in detail in Chapter 5. The reader is referred back to Chapter 5 for full details of the operation of these circuits.

Figure 7.9 shows the practical circuit of a simple full-wave motor-speed controller, using a triac as the control element. This circuit is similar to that shown in *Figure 5.16b*, except that the $L_1 - C_2$ r.f.-filter of *Figure 5.16b* is removed (since r.f. suppression is provided by the

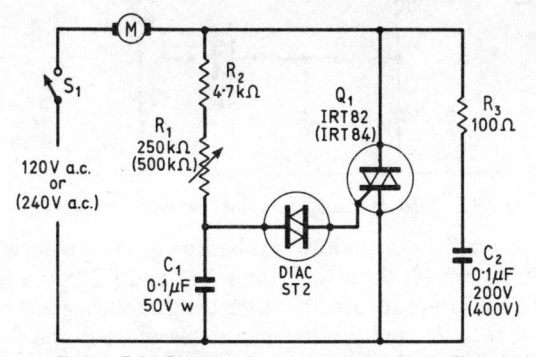

Figure 7.9. Simple triac motor-speed controller.

15 UNIVERSAL-MOTOR CONTROL PROJECTS

inductance of the motor), and the $R_3 - C_2$ rate-effect suppression network is wired across the triac main-terminals. Because of the inductive nature of the motor, the main-terminal currents and voltages of the triac are out of phase. Consequently, a high voltage switches across the triac terminals when the triac turns off at the end of each

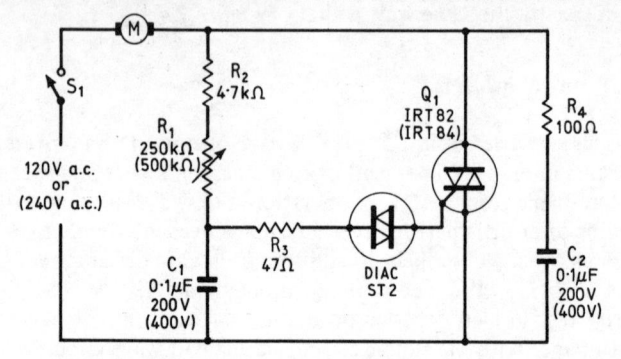

Figure 7.10. Improved triac motor-speed controller.

current half-cycle. R_3 and C_2 are used to limit the rate-of-rise of this switch-off voltage to a safe value, and thus prevent the triac from immediately being turned back on by rate effect.

As is pointed out in Chapter 5, circuits of the *Figure 7.9* type suffer from a 'backlash' effect, caused by the instantaneous voltage change that occurs across C_1 as the diac fires and turns on the triac in each

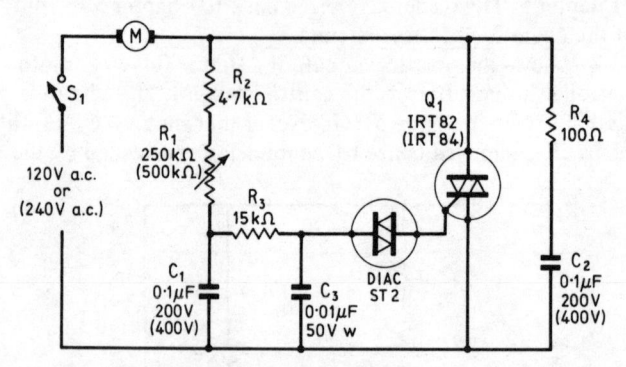

Figure 7.11. Triac motor-speed controller with gate-slaving.

half-cycle. *Figure 7.10* shows how this backlash can be reduced by wiring R_3 in series with the diac. *Figures 7.11* and *7.12* show how the backlash can be further reduced by using the gate-slaving technique. The *Figure 7.10, 7.11,* and *7.12* circuits are developed from the *Figure 5.17, 5.18,* and *5.19* circuits respectively. The four triac circuits

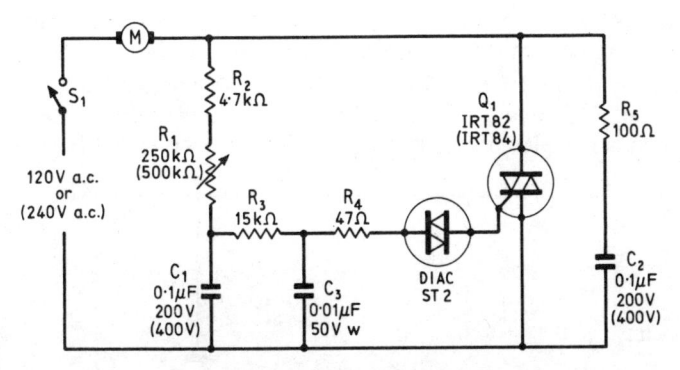

Figure 7.12. Minimum-backlash triac motor-speed controller.

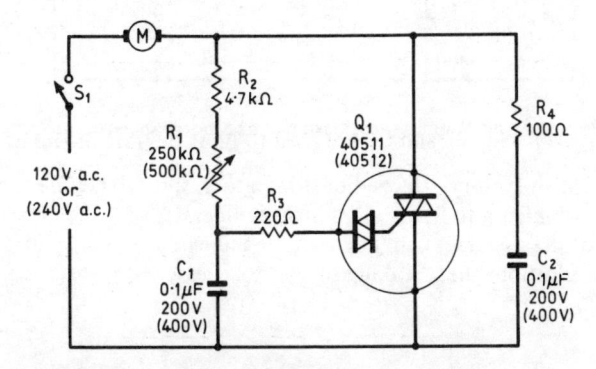

Figure 7.13. Simple quadrac motor-speed controller.

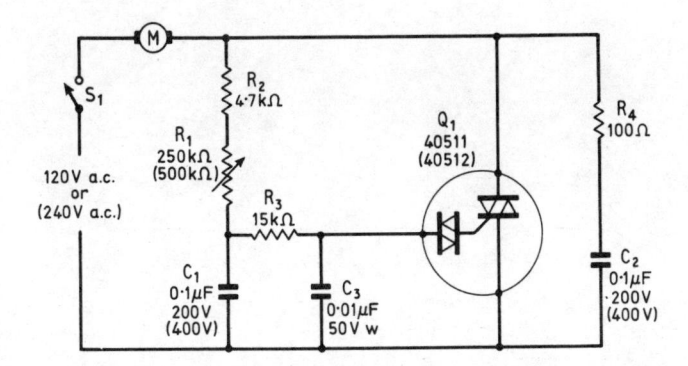

Figure 7.14. Quadrac motor-speed controller, with gate-slaving.

15 UNIVERSAL-MOTOR CONTROL PROJECTS

shown in *Figures 7.9* to *7.12* can each control motors with current ratings up to 8 A on 120 V lines, and up to 10 A on 240 V lines.

Finally, *Figures 7.13*, *7.14*, and *7.15* show how the *Figure 7.10*, *7.11*, and *7.12* circuits can be modified so that their individual triacs and diacs are replaced by a single quadrac. The specified quadracs are

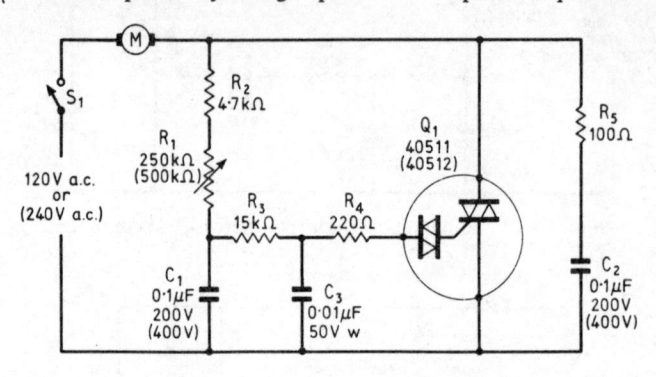

Figure 7.15. Minimum-backlash quadrac motor-speed controller.

supplied with factory-attached heat radiators, and can handle r.m.s. motor currents up to 2·2 A at ambient temperates of 25°C. The current ratings of the quadracs can, however, be increased to 6 A by attaching them to additional heat sinking so that their case temperatures are kept below 75°C.

CHAPTER 8

5 MISCELLANEOUS PROJECTS

S.C.R.s can be used in a number of useful projects that do not fit into any of the categories described under chapter headings already given in this volume. This final chapter describes five such projects. They comprise two 12 V self-regulating battery chargers, a model-train speed-controller, and two windshield-wiper pause-controllers for use in automobiles.

Self-regulating battery charger projects

Figure 8.1 shows the practical circuit of a self-regulating battery charger for use with 12 V batteries. The device charges the battery at a rate of 3 to 5 A when the accumulator is 'flat', but then automatically reduces the charge to a trickle rate when the battery reaches the fully charged state. The 0 to 4 A meter is an optional fitting. Circuit operation is as follows.

Transformer T_1 and bridge rectifier D_1 to D_4 step down and rectify the a.c. line voltage, and apply a charging current to the battery via $M_1 - R_1$ and SCR_1. The gate current of SCR_1 is derived from the rectified a.c. line via LP_1 and D_5. SCR_2 is wired between the $D_5 - LP_1$ junction and the lower supply line via R_6, and its gate current is derived from the battery via the $R_2 - R_3 - R_4$ potential divider and zener diode ZD_1. R_3 is adjusted so that SCR_2 triggers on only when the battery voltage reaches a 'fully-charged' value of 14·3 V.

Thus, when the battery is initially placed on charge its voltage is less than 14·3 V, and SCR_2 is off. Under this condition SCR_1 is triggered on via LP_1 and D_5 at the start of each half-cycle from the bridge rectifier, and a high charge current flows to the battery via $R_1 - M_1$ and SCR_1. R_1 limits the charge current to 3 or 4 amps over the approximate battery voltage range 10 to 14 V.

124 5 MISCELLANEOUS PROJECTS

As the battery charges up its terminal voltage rises in proportion to the state of charge and the charging current, and eventually reaches 14·3 V near full charge. At this stage SCR_2 is triggered on via ZD_1, and as it goes on it pulls the $D_5 - LP_1$ junction down towards the lower supply rail and thus removes the gate drive from SCR_1, which fails to turn on in the next half-cycle, thus removing the battery charge current. As SCR_2 goes on LP_1 lights up, giving a visual indication of circuit operation.

Once the charge current is removed the battery terminal voltage decays fairly rapidly back below the 14·3 V 'fully charged' value, and

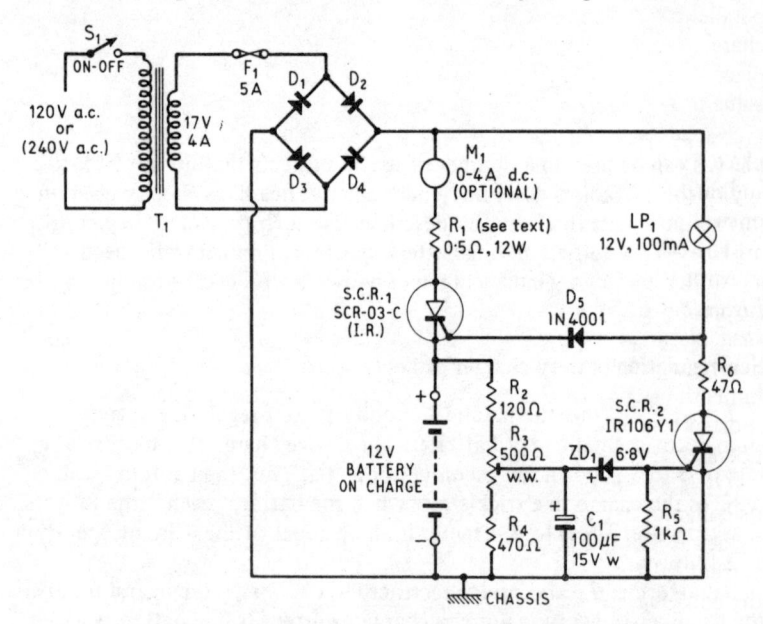

Figure 8.1. Self regulating 12 V battery charger.
D_1 to D_4 = 5 A or greater 30 p.i.v. silicon rectifiers.

SCR_2 turns off again and enables SCR_1 to turn on and apply more charging current to the battery. The decay time of the voltage increases in proportion to the state of battery charge. Consequently, when the battery first nears the fully charged state the decay period is short, and SCR_1 turns off for perhaps only one half-cycle in every ten, and little reduction in mean charge current takes place. The battery charge thus continues to rise, and the decay time slowly increases. When the battery is very near to full charge the decay time rises to such a value that SCR_1 fires on alternate half-cycles only, and the charge current falls to half of the maximum value.

Eventually, when the battery reaches full charge, the decay time becomes so long that SCR_1 turns on for perhaps only one half-cycle in every ten, and the charge current then falls to a 'trickle' rate of only a few hundred milliamps. Thus, the charge current does not fall sharply to zero as the battery reaches full charge, but reduces very slowly (over a period of several hours) from a full to a trickle value. The trickle charge is maintained indefinitely once the battery reaches full charge.

The *Figure 8.1* circuit presents few constructional or setting-up problems. R_1 has a nominal value of 0·5 ohms, and must have a 12 W or greater rating. This resistor can be made by series connecting two parallel pairs of 0·5 ohm 3 W resistors. R_1 is intended to limit the charge current to about 5 A when a 'dead' battery is connected in place. If the current is appreciably greater than this, increase the R_1 value to keep the current within limits.

R_3 can be set up by connecting a fully charged 12 V battery in place for two or three hours, and then VERY SLOWLY adjusting R_3 so that LP_1 starts to brighten and the charge current falls to one amp or so.

The only snag with the *Figure 8.1* circuit is that it is prone to fuse-blowing if the charger terminals are accidentally shorted together. *Figure 8.2* shows how the circuit can be modified to overcome this snag. Here, Q_1 is wired as a 4 A constant-current generator and is wired in series with the battery charging current path, and automatically limits the charge current to this value even if the charger terminals are shorted together. Lamp LP_2 lights up if a short occurs. The circuit operates as follows.

Q_1 is wired as an emitter follower, with emitter load R_6 and with its collector current derived from the battery charging network. D_6 and C_2 act as a simple voltage smoothing network that applies a steady current to $D_7 - D_8 - D_9$ and Q_1 base via R_7. A steady voltage equal to the combined forward volt drops of $D_7 - D_8$ and D_9 thus appears on Q_1 base. Due to the emitter follower action of Q_1 this steady voltage, minus the forward base-emitter volt drop of Q_1, is developed across R_6. This R_6 voltage is constant, and works out at about 1·2 V. Consequently, a constant current of about 4 A flows through R_6 via Q_1. This current flows into Q_1 collector from the battery charger circuit, and Q_1 collector thus acts as a constant-current source.

Normally, when a fully charged battery is connected in place, nearly all of the available charger voltage is developed across the battery. Under this condition very little voltage is developed between the emitter and collector of Q_1, which is nearly saturated. When, on the other hand, the battery charger terminals are shorted together, nearly all of the available voltage appears across Q_1 and none appears across the charger terminals. Lamp LP_2 is wired across Q_1, and thus lights up and gives a visual indication when a terminal short occurs.

When constructing the *Figure 8.2* circuit, note that Q_1 may dissipate some 50 or 60 W when the charger terminals are shorted

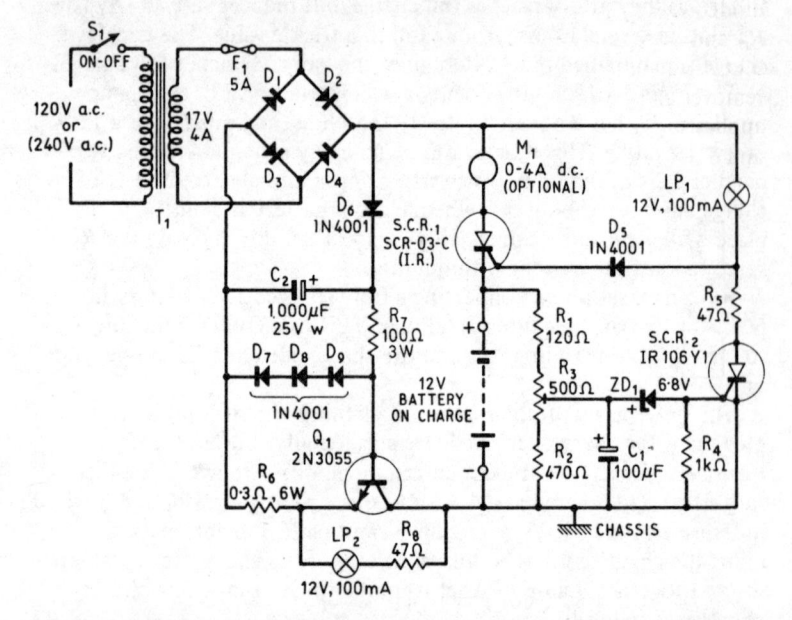

Figure 8.2. Self-regulating 12 V battery charger with short-circuit protection. D_1 to D_4 = 5 A or greater 30 p.i.v. silicon rectifiers.

together, so Q_1 must be connected to a large heat sink. The charger chassis can be used for this purpose, the metal body of Q_1 being bolted directly to the metal work (without the use of a mica washer) for maximum heat transfer. R_3 is set up in the same way as for the *Figure 8.1* circuit.

A model-train speed-controller

Figure 8.3 shows the practical circuit of a model-train speed-controller. The unit enables the speed of a 12 V model train to be varied smoothly from zero to maximum, with no sign of jerkiness. The maximum available output current is 1·5 A, and the unit incorporates an overload protection circuit that automatically limits the current to a safe value if a short-circuit occurs on the track. This feature eliminates the need for frequent fuse replacement and makes the controller virtually foolproof in use.

5 MISCELLANEOUS PROJECTS   127

Circuit operation is fairly simple. The line voltage is stepped down and full-wave rectified by T_1 and D_1 to D_4. At the start of each half-cycle of voltage the s.c.r. is off, and the voltage is thus applied to unijunction transistor Q_1 and to bipolar transistor Q_2 via R_1 and zener diode ZD_1. As soon as the half-cycle starts C_1 begins to charge up via R_7 and the parallel R_8-R_9 combination, and after a preset delay the u.j.t. fires and triggers the s.c.r. on. As the s.c.r. goes on it saturates and removes all power from Q_1 (which then resets) and from Q_2, and applies the remaining part of the half-cycle to the model train via R_2 and S_3.

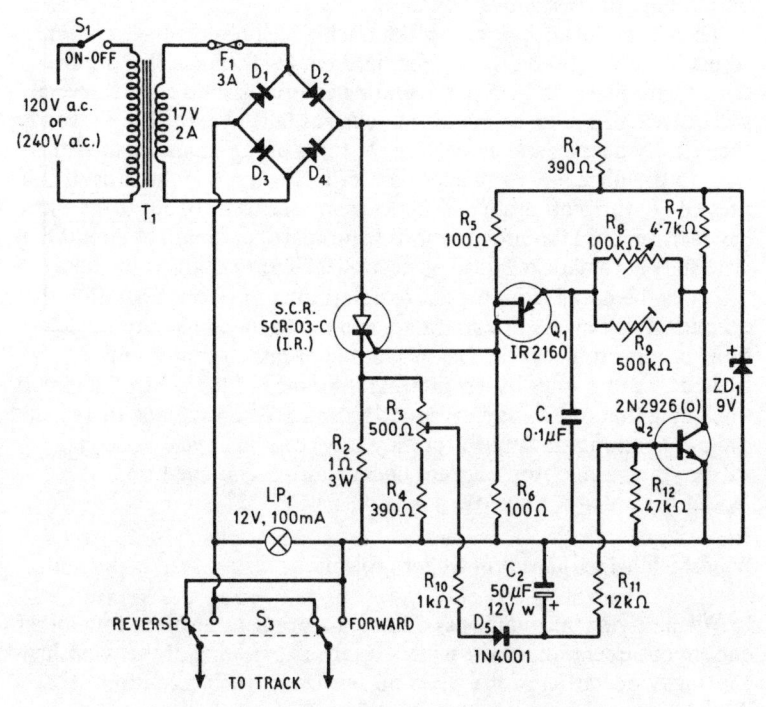

Figure 8.3. Model-train speed-controller with short-circuit protection. D_1 to D_4 = 3 A or greater 30 p.i.v. silicon rectifiers.

By adjusting R_8 the u.j.t. and s.c.r. can be made to trigger just after the start of each half cycle, or near the end of each half-cycle, or at any intermediate position. Phase-controlled s.c.r. triggering is thus obtained from the circuit, and the power (and the speed) of the train can be fully varied from zero to maximum via R_8. R_9 enables the range of R_8 to be preset so that the train speed just falls to zero at the maximum resistance setting of R_8. Lamp LP_1 is wired across the units output, and its intensity is proportional to the output power set by R_8. Switch S_3 enables the polarity of the output to be changed, to give forward and reverse of the train.

The overload protection part of the unit operates as follows. Each time the s.c.r. fires power is applied to the train via R_2, and a peak voltage is developed across R_2 and is proportional to the peak value of output current. Part of this voltage is tapped off via R_3 and is stored in C_2 via D_5, and the stored voltage is fed to the base of Q_2 via R_{11}. R_3 is adjusted so that this voltage is just sufficient to turn Q_2 on when the output current rises above 1·5 A.

Thus, if a short occurs across the track a high peak voltage appears across R_2 and C_2, and Q_2 is driven hard on. Q_2 then acts like a short circuit, and prevents Q_1 from operating and firing the s.c.r. for several successive half-cycles, so the output current falls to zero. The C_2 voltage then slowly decays, and eventually Q_2 turns off again and enables the s.c.r. to fire once more and apply power to the train. If the short is still present Q_2 turns off again and the current falls to zero, but if the short has been removed the circuit returns to normal operation. The circuit thus gives fully automatic overload protection if a short occurs on the track.

It may be noted from the above description of circuit operation that occasional half-cycles of current are applied to the units output when a short occurs on the track. The magnitude of this current is in fact limited to a few amps by the internal resistance of the controller circuit (including that of T_1), and causes no harm to the unit. Since these half-cycles of current are applied for only (say) one half-cycle in every fifteen, the mean output current under short-circuit conditions is in fact of the order of only 100 mA or so.

Windshield-wiper pause-controller projects

When driving in conditions of light rain or mist it is often found that continuous operation of the wipers results in smearing of the windshield and sticky operation of the wiper blades. Under such conditions it is desirable to operate the wipers for only a single stroke once every few seconds, rather than to operate them continuously. This 'pause-controlled' operation can be obtained automatically by using electronic control circuitry. Two such circuits are described in this final section of this volume.

These pause-control circuits can be added to about 65 per cent of modern automobiles with 12 V electrical systems, and enable the pause periods to be fully varied from roughly one to 20 seconds. The units are designed to operate in conjunction with conventional self-parking

5 MISCELLANEOUS PROJECTS 129

windshield wipers only. They can NOT be used with wipers fitted with dynamic braking facilities, or with wipers that are designed to self-reverse into the parked position.

Figure 8.4 shows the circuit of a conventional self-parking windshield wiper and its control switch. A self-parking switch (S_1) is built into the motor assembly and is activated by a cam that is mechanically coupled to the electric motor. This switch is wired between one side of the motor and ground, and is open when the wipers are parked, but closed when the wipers are sweeping. On-off switch S_2 is wired in parallel with S_1.

Thus, when S_2 is closed the motor is connected directly across the 12 V supply and the wipers operate; once the motor starts into its sweep S_1 is closed via the cam, so the motor continues to operate

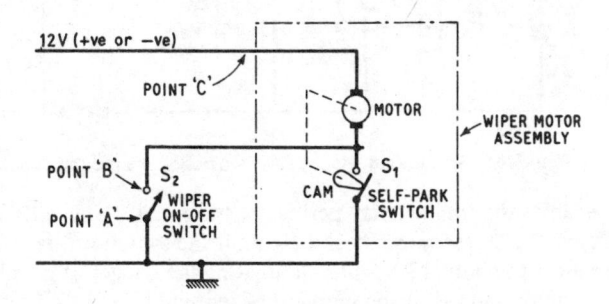

Figure 8.4. Circuit of conventional self-parking windshield wiper and its control switch.

independently of S_2, until eventually the motor returns to the self-park position and S_1 opens again. If S_2 is open at this moment the motor stops in the self-park position, but if S_2 is closed S_1 is shorted out and the wiper starts into its next sweep. Normal running currents of the motor are typically about 3 A, but switch-on surge currents may reach 24 A. When the motor switches off as S_1 opens the collapsing magnetic field of the motor causes a heavy back-e.m.f. (typically about 200 V peak) to be applied across S_1 and S_2.

Figure 8.5 shows the practical circuit of a negative ground pause-controller for use with wiper units of the type shown in *Figure 8.4*. The s.c.r. is connected in parallel with S_2, and is normally off. The circuit is activated by closing S_3. Circuit operation is fairly simple.

Assume that S_2 is open. When S_3 is closed voltage is applied to Q_1 via the wiper motor, and C_1 starts to charge up via R_1 and R_2. Eventually, after a preset delay, C_1 reaches the firing potential of Q_1, which then fires and triggers the s.c.r. on. Once the s.c.r. is triggered it self-latches and acts like a closed switch, so power is applied to the motor and the wipers start to move. As the wipers move into the sweep self-park switch

S_1 closes and shorts out the s.c.r. and Q_1 circuit, which turn off and reset, and the motor completes the sweep via S_1. At the end of the sweep S_1 again opens, so Q_1 starts into another timing cycle, at the end of which it again fires the s.c.r. and restarts the wipers. This process continues so long as S_3 is closed. When S_3 is opened power is no longer applied to Q_1 at the end of each sweep, so circuit operation ceases.

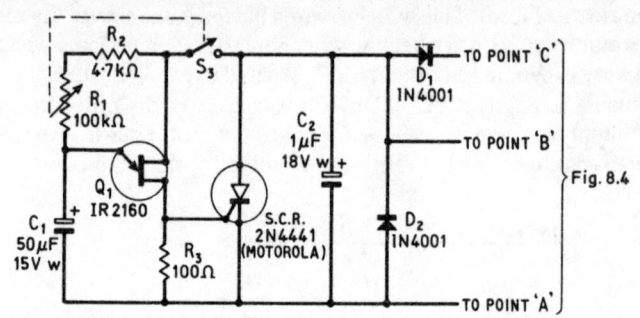

Figure 8.5. Windshield-wiper pause-controller, -ve ground version.

R_1 enables the timing delay periods of the circuit to be varied from roughly one to 20 seconds. D_1 and D_2 limit the switch-off back-e.m.f. of the motor to about 13 V, thus eliminating the danger of possible semiconductor damage at the moment of switch-off, and C_2 limits the rate-of-rise of the s.c.r. anode voltage, thus ensuring that the s.c.r. is not turned on by switching transients.

The positive ground version of the unit, shown in *Figure 8.6*, is almost identical to that described above, except that the circuit connections are changed. In both circuits the specified s.c.r. has an 8 A average and 80 A surge rating, and thus has ample safety margins for operating the wiper motors. The s.c.r. should be mounted on a small heat sink in the completed circuits, to prevent overheating.

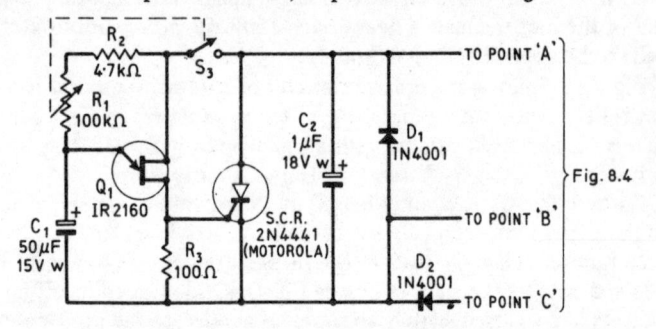

Figure 8.6. Windshield-wiper pause controller, +ve ground version.

Before starting construction, make sure that the wipers are suitable for operation by the selected pause-controller circuit. This can be determined very easily, as follows. Gain access to the vehicle's wiper control switch terminals via a couple of clip-on wires. Switch on the ignition, and touch-connect a 5 A to 8 A SLOW-BLOW fuse across the switch terminals. The wipers should operate in the normal way. Let the wipers work through two or three sweeps, and then remove the fuse half way through a following sweep. The wipers should continue through this sweep and then stop in the self-park position. If the wipers work as described, the electronic unit can safely be used on the vehicle. If, on the other hand, the fuse blows during these tests, or the wipers fail to self-park, the unit can not be used on the vehicle.

APPENDIX

TABLE 9.1. Classification data for semiconductor devices used in this volume.

Device No.	Device Type	Manufacturer	Outline
2N2926 (o)	npn Transistor	General Electric	1
2N3702	pnp Transistor	Texas	2
2N3704	npn Transistor	Texas	2
2N3055	npn Transistor	R.C.A.	3
IR2160	Unijunction Transistor	International Rectifier	4
IR106Y1	S.C.R.	International Rectifier	5
SCR − 01 − C	S.C.R.	International Rectifier	6
SCR − 03 − C	S.C.R.	International Rectifier	6
2N3228	S.C.R.	R.C.A.	7
2N3525	S.C.R.	R.C.A.	7
2N4441	S.C.R.	Motorola	8
ST2	Diac	General Electric	None (This device is non-polarised)
IRT82	Triac	International Rectifier	9
IRT84	Triac	International Rectifier	10
40575	Triac	R.C.A.	11
40511	Quadrac	R.C.A.	12
40512	Quadrac	R.C.A.	12
2N4990	S.U.S.	General Electric	13
2N4991	S.B.S.	General Electric	14
PA424	Integrated Circuit	General Electric	15

Note: In case of difficulty, U.K. readers can (at the time of writing) obtain these semiconductor devices, except for the IRT82, from Arrow Electronics Ltd., 7 Coptfold Road, Brentwood, Essex

Appendix 133

TABLE 9.2. Semiconductor outlines.

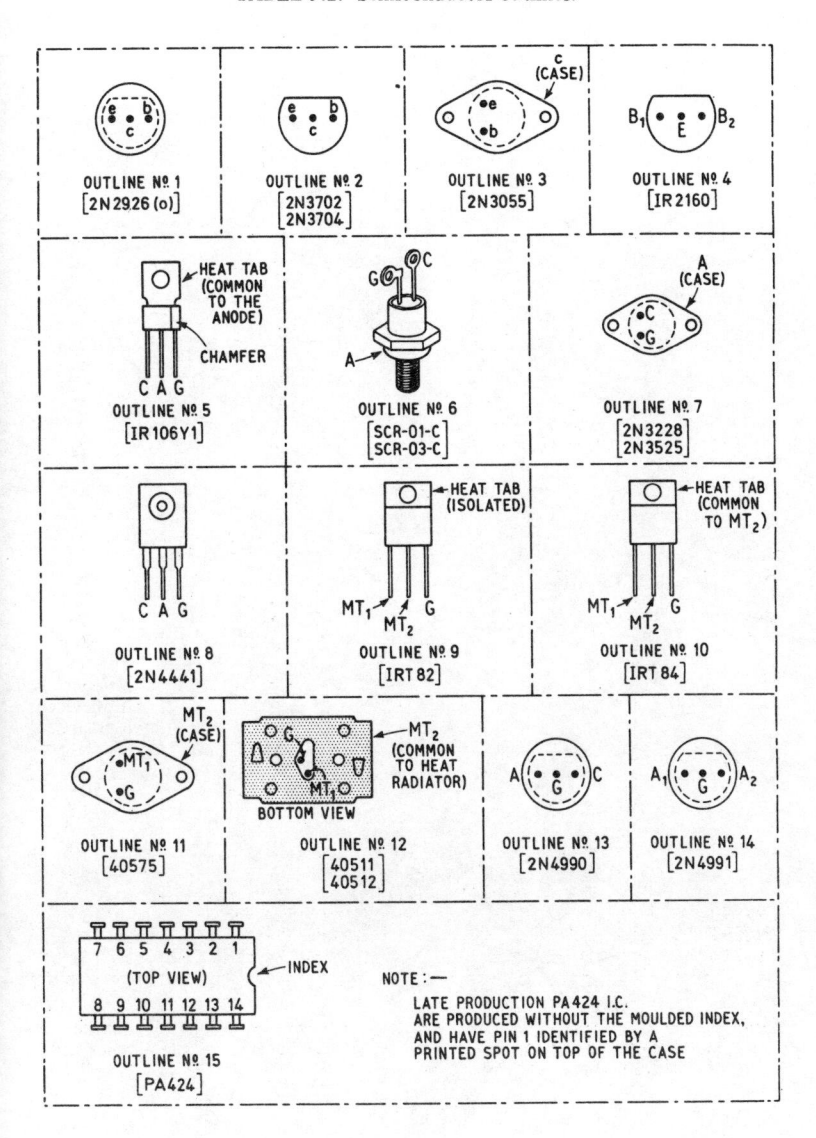

OUTLINE No. 1	[2N2926 (o)]
OUTLINE No. 2	[2N3702, 2N3704]
OUTLINE No. 3	[2N3055]
OUTLINE No. 4	[IR2160]
OUTLINE No. 5	[IR106Y1]
OUTLINE No. 6	[SCR-01-C, SCR-03-C]
OUTLINE No. 7	[2N3228, 2N3525]
OUTLINE No. 8	[2N4441]
OUTLINE No. 9	[IRT82]
OUTLINE No. 10	[IRT84]
OUTLINE No. 11	[40575]
OUTLINE No. 12	[40511, 40512]
OUTLINE No. 13	[2N4990]
OUTLINE No. 14	[2N4991]
OUTLINE No. 15	[PA424]

NOTE:— LATE PRODUCTION PA424 I.C. ARE PRODUCED WITHOUT THE MOULDED INDEX, AND HAVE PIN 1 IDENTIFIED BY A PRINTED SPOT ON TOP OF THE CASE

Index

A.C. circuits, 8
   half-wave on-off, 8
   full-wave control, 9
A.C. lamp flasher projects, 77, 78
   automatic, 79
      synchronous, 81
A.C. time-delay projects, 62
A.C. power switch
   with s.b.s. aided triggering, 24
   with transistor-aided d.c. triggering, 25
   with u.j.t. triggering, 26
A.C. power switching projects, 23
Alarm projects
   burglar, 39
      self-latching, 42
      tamper-proof, 41
   contact-operated, 38
   electronic, 28
   fire, 39, 52
   frost, 52
   light-beam, 47
      unbeatable, 49
   light-operated, 46
   non-latching, 7
   over-temperature, 52, 53
   remote-operated, 39
   self-latching, 7
   smoke-operated, 50
   steam-operated, 44
   temperature-operated, 51
   under-temperature, 52
   water-operated, 44
Automatic integral-cycle control projects, 105
Automatic light-activated a.c. lamp projects, 68
   indoor use, 70
   outdoor use, 69
Automatic overload-switching projects, 35
Automatic turn-off d.c. switch, 58

Basic a.c. circuits, 8

Basic d.c. circuits, 4
   s.c.r., 4
Basic d.c. switch
   s.c.r., 2
   triac, 12
Basic principles
   s.c.r., 1
   triac, 11
Basic theory, s.c.r., 1
Battery-charger projects, self-regulating, 123
Bistable, s.c.r., 6
Burglar alarm, 39

Cadmium-sulphide photocell, 46
Capacitor turn-off circuit, 5
Characteristics
   s.c.r., 2
   universal-motor, 110
Classification data, 132
Contact-operated alarm projects, 38

Data, classification, 132
D.C. circuits, s.c.r., 4
D.C. delayed turn-on projects, 82
D.C. lamp-flasher projects, 74
D.C. light-activated lamp flasher, 75
D.C. on-off circuits, 4
   s.c.r., 4
Delay projects, 54
   a.c., 62
   d.c., 58
Delayed turn-on line switch, 64
Diac, 20
   ST2, 86, 119, 132, 133

Electronic alarm projects, 38

Fire alarm, 39, 52
Frost alarm, 52
Full-wave a.c. control circuit, 9

136 INDEX

Full-wave control projects, 119
Full-wave line switch, 11
Full-wave s.c.r. on-off circuit
  with a.c. load, 10
  with d.c. load, 10

Gate sensitivity, triac, 12

Half-wave control projects, 111
Half-wave motor-speed controller, 112
  with self-regulation, 114
    high-performance, 114
  with transistor trigger, 113
Half-wave universal-motor control projects, 110
Heater control projects, 94
Heater controller, self-regulating, 107
Heater, thermistor-controlled, 95, 97
  synchronous, 97, 99, 101
  with tri-mode operation, 96
High performance lamp dimmer, 92

Integral-cycle heater controller, self-regulating, 107
Integral-cycle variable power projects, 100
  automatic, 105
Interference, radio frequency, 16
IR106Y1, 38, 126
IR2160, 26, 54, 127, 132, 133
IRT82, 23, 94, 132, 133
IRT84, 23, 94, 132, 133
Isolated-input line switch, 27

Lamp-chaser projects
  d.c., 82
  repetitive, 83
Lamp-control projects, 68
  time-delayed, 71
Lamp-control unit, 91
Lamp-dimmer projects, 85
  high-performance, 92
  phase-triggered, 85
  simple quadrac, 89
  single-lamp, 87
  with gate slaving, 88
Lamp-flasher projects
  a.c., 77

Lamp-flasher projects (contd)
  automatic, 79
    synchronous, 81
  d.c., 74
  light-activated, 75
  twin-lamp d.c., 76
L.A.S.C.R., 18
Light-activated a.c. lamp projects, 68
  indoor use, 70
  outdoor use, 69
Light-activated alarm, 46
Light-activated s.c.r., 18
Light-beam alarm, 47
Light-dependent resistor, 46
Light-operated alarms, 46
Line-triggered a.c. power switch, 23

Miscellaneous projects, 123
Miscellaneous thyristor and trigger devices, 18
Model-train speed-controller, 126
Motor-speed controller, half-wave, 112
  high-performance, 114
  with self-regulation, 114
  with transistor trigger, 113
Motor-speed controller, quadrac, 121
  with gate slaving, 121
Motor-speed controller, triac, 119
  with gate slaving, 120

On-off circuits
  automatic overload switch, 35
  automatic turn-off, d.c., 58
  basic d.c. switch, s.c.r., 2
    triac, 12
  d.c. triggered line switch, 13
  delayed turn-on, 54
    line, 64
  full-wave line switch, 11
  full-wave with d.c. load, 10
  half-wave a.c., 8
  half-wave line driven, 9
  isolated-input line switch, 27
  push-button self-latching line switch, 29
  steam-operated, 27, 28
  synchronous, 97
  water-operated, 27, 28
Oscillator, relaxation, u.j.t., 21
Outlines, 133

INDEX

Over-temperature alarm, 52

PA424, 32, 100, 104, 109, 132, 133
Phase-triggered power-control
   principles, 15
   waveforms, 15
Photocell, cadmium sulphide, 46
Power-control projects, 100
   automatic, 105
Power switching, a.c., 23
Precision over-temperature alarm, 53
Push-button operated self-latching line
   switch, 29

Quadrac, 20
   motor-speed controller, 121
   with gate slaving, 121
   40511, 89, 121, 122, 132, 133
   40512, 89, 121, 122, 132, 133

Radio-frequency interference, 16
Rate effect demonstration circuit, 8
Relaxation oscillator, 21
   waveforms, 22
Remote-operated alarm, 39
Repetitive lamp chaser, 83

S.B.S., 19
   symbol, 20
   2N4991, 24, 132, 133
S.C.R.
   basic a.c. circuits, 8
   basic d.c. circuits, 4
   basic d.c. switch, 2
   basic principles and projects, 1
   basic theory, 1
   bistable, 6
   capacitor turn-off circuit, 5
   characteristics, 2
   d.c. on-off circuit, 4
   full-wave a.c. control, 9
   full-wave line switch, 11
   full-wave on-off circuit with a.c.
      load, 10
      with d.c. load, 10
   half-wave a.c. on-off circuit, 8
   IR106Y1, 4, 38, 54, 124, 126,
      132, 133

S.C.R. (contd)
   line-driven half-wave on-off
      circuit, 9
   non-latching alarm circuit, 7
   rate-effect demonstration
      circuit, 8
   symbol, 2
   transistor equivalent, 2
   2N3228, 9, 112, 132, 133
   2N3525, 9, 112, 132, 133
   2N4441, 130, 132, 133
   SCR-03-C, 124, 126, 132, 133
S.C.S., 18
   symbol, 18
Self-latching alarm circuit, 7
Self-latching burglar alarm, 42
Self-regulating battery-charger
   projects, 123
Semiconductor
   classification data, 132
   outlines, 133
Sensitivity, triac gate, 12
Silicon bilateral switch, 19
Silicon controlled switch, 18
   symbol, 18
Silicon unilateral switch, 19
   symbol, 19
Simple lamp dimmer, 87
Simple quadrac lamp dimmer, 89
Smoke-operated alarm, 50
Speed-controller, model train, 126
Steam-operated alarms, 44
Steam-operated line switch, 27, 28
S.U.S.
   symbol, 19
   2N4990, 112, 132, 133
Switch
   automatic overload, 35
   d.c. triggered line, 13
   full-wave line, 11
   isolated-input line, 27
   line-triggered, 13, 14
      a.c. power, 33
   push-button self-latching line, 29
   s.c.r. basic, 2
   silicon bilateral, 19
   silicon controlled, 18
   silicon unilateral, 19
   steam-operated, 27, 28
   synchronous, 29, 30, 34
   three-way line, 14
   triac, 12
   water-operated, 27, 28

138 INDEX

Symbols
    diac, 20
    l.a.s.c.r., 18
    quadrac, 20
    s.b.s., 20
    s.c.r., 2
    s.c.s., 18
    s.u.s., 19
    triac, 11
    unijunction transistor, 21
Synchronous on-off control projects, 97
Synchronous power-switching projects, 29
    zero-switching, 30
    using i.c., 34
Synchronous thermostat-controlled heaters, 97

Tamper-proof burglar alarm, 41
Temperature-operated alarms, 51
Thermistor-controlled heater, 95
    synchronous, 97, 99, 101
    with tri-mode operation, 96
Three-way line switch, 14
Thyristor devices, miscellaneous, 18
Time-delay projects, 54
Time-delayed lamp-control projects, 71
Transistor equivalent of s.c.r., 2
Triac
    basic principles and projects, 11
    basic switch with d.c. gate drive, 11
    d.c. triggered line switch, 13
    gate sensitivity, 12
    IRT82, 13, 62, 94, 95, 132, 133

Triac (contd)
    IRT84, 13, 62, 94, 95, 132, 133
    line-triggered line switch, 13, 14
    motor-speed controller, 119
        with gate slaving, 120
    phase-triggered power control principles, 15
    symbol, 11
    three-way line switch, 14
    waveforms, 15
    40575, 94, 95, 132, 133
Trigger devices, miscellaneous, 18
Twin-lamp d.c. flasher, 76

Unbeatable light-beam alarm, 49
Under-temperature alarm, 52
Unijunction transistor, 21
    IR2160, 26, 54, 127, 132, 133
    Simple relaxation oscillator, 21
    waveforms, 22
    symbol, 21
Universal-motor characteristics, 110
Universal-motor control projects, 110
    half-wave, 111

Water-operated alarms, 44
Water or steam-operated line switch, 27, 28
Windshield wiper pause controller projects, 128

Zero-switching synchronous overload switch,
    self-latching, 36